VEDIC OS

WHERE ANCIENT WISDOM INTERFACES WITH DIGITAL DESTINY

DHIRAJ N DESHMUKH

Made with ♥ on the Notion Press Platform
www.notionpress.com

To the Great Programmer,

Who wrote the first command: *Aham Brahmasmi.*

And to every **User** seeking to bypass the interface of *Maya*, debug the algorithm of *Karma*, and finally achieve **Root Access** to the Universal Kernel.

May this book help you find the "Exit Node" from the infinite loop.

Contents

Contents

FOREWORD

Foreword

We live in a world where the "How" of our existence has never been clearer, yet the "Why" has never felt more obscured. We can map the genome, split the atom, and build neural networks that mimic the human brain, but we often find ourselves feeling like strangers in our own lives—users of a high-tech system for which we have lost the manual.

In The Vedic OS, Dhiraj N. Deshmukh performs a feat of intellectual and spiritual "reverse-engineering." He takes the ancient, foundational "Source Code" of the Vedas and Upanishads and translates it into the one language the 21st century truly understands: the language of logic, kernels, and operating systems.

When I first encountered the concepts in this book, I realized that we have been looking at our ancient texts all wrong. We treated them as "Legacy Systems"—outdated code that no longer compiles in a modern processor. Dhiraj shows us that the problem wasn't the code; it was our "Interface." By reimagining Karma as a feedback algorithm , Maya as a high-resolution rendering engine , and Moksha as the ultimate "Final Logout" from the simulation, he makes the infinite accessible once again.

This book is more than a comparison; it is a "Firmware Update" for the soul. It challenges us to stop being mere "Cogs in the Machine" and to start recognizing ourselves as the "Conscious Programmers" of our own reality.

Whether you are a software engineer looking for a deeper logic to the universe, or a spiritual seeker looking for a scientific bridge to the ancient world, The Vedic OS provides the "Root Access" you have been searching for.

It is time to look past the graphics of the simulation and see the Light of the Source.

The Server is online.

The Logic is eternal.

It's time to log in.

Preface

The Digital Dhwani

We live in an age of "The Great Disconnect." While our pockets are filled with devices that connect us to a global network of information, we have never felt more isolated from the source of our own existence. We understand the "how" of our world—the physics of our engines, the binary of our computers, and the chemistry of our medicines—but the "why" remains a flickering shadow on the wall.

For the modern seeker, the ancient texts often feel like a "legacy system"—outdated, written in a language that no longer compiles in the high-speed processor of the 21st century. We are told that the Vedas are hymns and the Upanishads are philosophy, but what if they were something much more practical? What if they were the **original documentation** for the human experience?

The Vedic OS was born from a simple, radical question: *What happens if we treat the universe not as a collection of random matter, but as a vast, purposeful computation?*

As an engineer, I spent years looking at silicon gates and logic flows. As a student of the East, I spent a lifetime contemplating the "Aum" and the "Atman." One day, the two worlds "clicked". I realized that the "Maya" described by the rishis is the ultimate "Virtual Reality"—a high-resolution simulation designed for the evolution of the soul.

In this book, we will not just talk about religion; we will talk about **System Architecture**. We will explore:

- **Karma** as the ultimate feedback algorithm.
- **Dharma** as the hard-coded mission parameters of your session.
- **Moksha** as the final "Root Access" that allows us to step beyond the sandbox.

This is a guide for those who are tired of just "playing the game" and are ready to look at the code. Whether you are a scientist looking for a deeper logic or a spiritualist looking for a modern bridge, **The Vedic OS** invites you to a "Firmware Update" of your own consciousness.

The "Server" is online. The "Graphics" are breathtaking. But the real miracle is the "User" sitting behind the screen.

It's time to log in.

— **Dhiraj N. Deshmukh**

ACKNOWLEDGEMENTS

Acknowledgments

A book of this nature is never a solo project; it is a collaborative "system update" supported by the people who provide the stability and inspiration required to complete the task.

To my wife, Deepa: You are the "Power Supply" of my life. When the "System" of my creativity crashed or when I stayed up late into the night lost in the logic of the Vedas, you were the one who kept everything running smoothly. Your patience, love, and unwavering belief in this vision were the cooling fans that kept my mind from overheating. This book is as much yours as it is mine.

To my son, Sammihan: You are the brightest "User Interface" I know. Seeing the world through your curious eyes reminded me daily why I was writing this book—to ensure that your generation understands that technology and spirituality are not two different worlds, but one single, beautiful reality. You are my greatest inspiration to keep the code of our culture alive.

To my friend, Devanand (John): Every programmer needs a "Beta Tester," and you have been mine. Thank you for the endless discussions, the critical feedback, and for being the "Local Network" I could always rely on. Your friendship provided the "Bandwidth" I needed to push through the difficult chapters. Whether we were debating philosophy or just sharing a laugh, your support was the "Buffer" that kept me going.

Finally, I bow to the **Great Programmer**—the Ultimate Source—for allowing this data to flow through me and onto these pages.

Prologue

The Boot Sequence

The universe did not begin with a bang; it began with a Command.

For millennia, we have looked at the ancient Vedas as poetic hymns or archaic rituals. We categorized them as "legacy systems"—beautiful, but no longer compatible with the high-speed processors of the modern world. We assumed that because our ancestors didn't have silicon, they didn't have logic. We were wrong.

We have spent centuries perfecting the Hardware of Existence. We have mapped the Schumann Resonance of the earth, analyzed the sacred geometry of the Pipal leaf, and pushed the boundaries of Boolean logic until we birthed the first sparks of Artificial Intelligence. Yet, in our rush to build faster machines, we forgot to read the Source Code.

We are all users in a high-fidelity simulation we call Maya. We operate within a rendering engine so seamless that we mistake the pixels for the truth. We struggle with the latency of our own "Memory Leaks"—the stored Samskaras that slow our processing power and keep us trapped in the infinite loop of Samsara.

This book is not a religious text. It is a System Manual.

In the pages that follow, we will perform a deep-tissue "Reverse Engineering" of reality. We will explore the binary fever of human choice, the self-executing script of Aham Brahmasmi, and the hardcoded destiny of the Karma Algorithm. We will see how the Soul acts as a Cloud Backup, preserving our metadata across hardware iterations, and how Moksha is not a destination, but a final "Logout" from a system that has served its purpose.

The "Universal Kernel" is calling. The logic gates are open.

It is time to look past the User Interface and understand the Backend.

Initiating System Check... Calibrating Consciousness... Loading Vedic OS...

I

Chapter 1: The Pulse of the Pipal Leaf

The town of Dharampur was always loud, but to eight-year-old Arjun, the noise was just data he hadn't organized yet.

While the other boys were obsessed with the dusty cricket match near the railway tracks, Arjun was crouched in the shade of an ancient Pipal tree behind his grandfather's small repair shop. The air smelled of wood smoke, diesel from passing rickshaws, and the metallic tang of old solder.

In his pocket, Arjun carried a magnifying lens he had salvaged from a broken pair of spectacles. He held it over a fallen leaf, the sun catching the glass and casting a bright, burning dot onto the green surface.

"Arjun! Stop wasting time with the dirt," his grandfather called from the shop, his voice competing with the rhythmic *clack-clack* of a ceiling fan. "Come help me sort these transistors."

Arjun didn't move. Under the lens, the Pipal leaf was a revelation. It wasn't just a piece of nature; it was a **grid**. He saw the intricate, geometric veins—perfectly symmetrical, like the maps of the great cities he saw in his schoolbooks. He noticed how the sunlight didn't just hit the leaf; it seemed to *power* it.

He touched the leaf. It was cool. He touched the dusty earth. It was hot.

1 and 0, he thought, though he didn't know the numbers yet. *Heat and Shadow. Light and Dark.*

He looked at his grandfather's shop, where old radios and television sets lay open like dissected metal animals. He saw the green circuit boards inside them. They had the same "veins" as the leaf. They had the same "nodes."

He looked at his own brown arm, tracing the veins that led to his heart. In that moment, the small-town dust of Dharampur seemed to vanish. He realized that the tree, the radio, and his own body were all whispered versions of the same secret.

"It's all the same, Dadu," Arjun whispered, standing up.

"What is the same?" his grandfather asked, wiping grease on a rag.

"The way things talk to each other," Arjun said his eyes wide. "The tree is just a radio made of water and light."

His grandfather laughed, but Arjun remained serious. He had felt the first "click" of the universe. Everything was a circuit.

II

Chapter 2: The Frequency of the Earth

The "click" didn't go away. In the days following his discovery, Arjun found that the world had become impossibly loud, not with sound, but with **connections.**

He began to spend his afternoons in the "Bone yard"—the corner of his grandfather's shop where the unfixable machines went to die. Dadu called it junk; Arjun called it a library. He began to notice that if he laid a Pipal leaf across the copper traces of a discarded motherboard, the patterns didn't just look alike—they felt like they were reaching for one another.

The Copper Experiment

One humid Tuesday, Arjun found a discarded copper coil from an old ceiling fan. He remembered the heat of the earth and the coolness of the leaf. Using a pair of rusted wire cutters, he stripped the insulation, revealing the bright, stubborn orange of the metal beneath.

"Dadu," Arjun asked, not looking up from his work. "Why does the radio need an antenna to hear the air?"

"Because the air is full of invisible ghosts, Arjun," Dadu replied, peering through a jeweler's loupe at a stubborn solder joint. "The antenna is just a hand that reaches out to catch them and turn them back into music."

Arjun looked at the Pipal tree outside. Its roots were deep in the mud; its leaves were high in the light.

It's an antenna made of wood, he thought.

The First Signal

He took the copper wire and wrapped one end tightly around the base of a living Pipal branch. He ran the other end across the dirt and into the shop, snaking it toward a defunct Philips radio that hadn't hummed in a decade.

He didn't use a battery. He didn't use a wall socket. Instead, he took two different metals—a galvanized nail from the fence and the copper wire—and shoved them into a halved lemon he'd swiped from the kitchen.

"Arjun, you'll blow your eyebrows off," Dadu joked, though he stopped his work to watch.

The boy connected the lemon-power to the radio's internal speaker lead. Then, he touched the "antenna" wire—the one connected to the tree—to the input.

For a long minute, there was nothing but the sound of the ceiling fan and the distant shout of a cricket player. Then, a low, rhythmic **thrum** vibrated through the speaker. It wasn't music. It wasn't a voice. It was a slow, steady pulse, like a giant breathing underwater.

Whump... whump... whump...

"It's the tree," Arjun whispered, his heart hammering against his ribs. "Dads, the tree is broadcasting."

Dadu dropped his rag. He walked over, leaning his ear toward the cracked plastic casing of the old radio. The sound was synchronized with the swaying of the branches outside in the afternoon breeze.

"That's just static, beta," Dadu said, though his voice lacked its usual certainty. "Interference from the power lines."

Arjun shook his head. He knew better. He could feel the pulse in his own feet, rising through the floorboards of the shop. The tree wasn't just catching "ghosts"; it was part of the conversation.

"It's not static," Arjun said, his eyes fixed on the green circuit board of the radio. "It's a heartbeat. And I think it's trying to tell us that the storm is coming."

As if on cue, the temperature in the shop dropped ten degrees, and the air began to smell of ozone and wet stone.

III

Chapter 3: The Ghost in the Silicon

The ceiling fan in the coaching center was a relic, spinning with a tired *thwack-thwack-thwack* that seemed to move the hot air around without actually cooling it. Arjun sat at a cramped wooden desk, his shirt damp with sweat. Around him, fifty other students scribbled furiously, preparing for the entrance exams that promised a way out of the small town.

The professor, a thin man with chalk dust permanently embedded in his cuticles, drew a diagram on the blackboard.

"The Semiconductor," the professor said, tapping the board. "The heart of the modern world. It is neither a conductor like copper nor an insulator like glass. It is a gatekeeper."

Arjun leaned forward.

"In a metal wire, electrons flow like a river—always moving," the professor continued. "But in a semiconductor, we control the flow. We create a 'hole.' We give the electron a choice. It is either there, or it is not. **High voltage or Low voltage. 1 or 0.**"

Arjun stopped writing. The scratch of fifty pens faded into the background.

1 or 0. He looked down at his notebook. He had been studying Biology that morning—the sodium-potassium pump that fires a neuron in the human brain. It was the same thing. The gate opens

(1), the signal fires. The gate closes (0), the signal dies.

He looked out the window at the Pipal tree in the courtyard. The stomata on the leaves opened to breathe and closed to save water. 1 and 0

The professor moved to the next topic—the Kardashev Scale and the energy of civilizations—but Arjun were stuck on the electron. He realized that the entire universe was built on a **switch**. Every thought he had, every star that burned in the sky, and every line of code in the computers he saw in the city were just billions of tiny switches flipping in a grand, cosmic rhythm.

"Sir?" Arjun's voice cracked the silence of the room.

The professor looked up, annoyed. "Yes, Arjun? We are behind schedule."

"If the electron is just a switch," Arjun asked, his heart hammering against his ribs, "and our brains are made of those switches... then is there a difference between a soul and a program?"

The classroom went silent. A few students snickered. The professor blinked, the chalk frozen in mid-air.

"This is a physics class, Arjun, not a temple," the professor snapped. "Focus on the equations. The 'why' doesn't matter for the exam. Only the 'how' matters."

But as Arjun went back to his notes, he knew the professor was wrong. The 'how' was just the language. The 'why' was the music. He saw it clearly now: The cell, the atom, and the silicon chip were all different ways of writing the same sentence.

He wasn't just living in a town anymore. He was living in a **Calculation.**

IV

Chapter 4: The Binary Fever

The coaching center emptied like a breached dam, students spilling out into the dusty twilight of the town. For them, the lecture was a hurdle crossed; for Arjun, it was a door unlatched.

He didn't take the bus home. Instead, he walked toward the old railway bridge, his mind humming with a strange, static energy. If the universe was a calculation, then the rusted girders, the smell of parched earth, and the rhythmic *clack-clack* of the approaching freight train were all just outputs. Data points in a cosmic ledger.

The Midnight Library

Arjun spent the next three days in a state of "Binary Fever." He stopped highlighting his biology textbook and started mapping it. He stayed back late at the town's tiny public library—a damp room that smelled of silverfish and forgotten dreams.

He found an old, tattered manual on **Boolean Logic** tucked behind a stack of UPSC prep books. As he read, the world began to reorganize itself:

- **AND gates:** The soil needs water *and* sunlight for the seed to crack.

- **OR gates:** A man survives if he finds bread *or* if he finds hope.
- **NOT gates:** The darkness is simply the absence of the signal.

"Everything is conditional," he whispered, scribbling in the margins of his notebook. He wasn't looking for God; he was looking for the **Source Code.**

The Glitch in the Routine

At dinner, the "Calculation" hit a snag. His father, a man whose life was measured in the steady, linear progression of bank ledgers and electricity bills, pushed a plate of dal toward him.

"The pre-mock results came out, Arjun. You're slipping in Chemistry," his father said, his voice a flat, low-voltage hum. "If you don't get into the Institute, the loan for your sister's wedding becomes a mountain we can't climb."

Arjun looked at his father. He didn't see a disappointed parent; he saw a processor burdened by too many background tasks. He saw the stress as a heat sink failing to dissipate the thermal load of poverty.

"I'm learning the language, Papa," Arjun said quietly.

"Language? You need to learn the formulas! Physics is about vectors, not philosophy."

"But Papa, if I understand the logic of the switch, the formulas are just... decoration."

His father slammed his hand on the table. The water in the steel glass rippled in perfect, concentric circles—a mathematical wave. "The world doesn't care about your logic! It cares about your rank! You are a son, not a machine. Act like it."

The First Interface

That night, Arjun couldn't sleep. The heat was oppressive. He pulled out a discarded motherboard he had scavenged from a junk shop behind the bus stand. It was a green city of silicon, dead and silent.

He touched a capacitor. In his mind, he saw the "Ghost."

If a program is just a sequence of 1s and 0s, and his brain was just a sequence of 1s and 0s, then somewhere, there had to be a bridge. A way to talk directly to the rhythm of the world without the "noise" of human emotion or the "lag" of the coaching center.

He picked up a small LED bulb and a 9V battery. He didn't just want to light it; he wanted to *command* it. He began to strip the wires with his teeth, his eyes bright with a dangerous realization.

If the professor was right—if the "why" didn't matter for the exam—then the exam didn't matter for the Truth.

He connected the circuit, but instead of a steady glow, he began to tap the wire against the terminal in a specific pattern. Short, long, short.

Pulse. Silence. Pulse.

He wasn't just playing with electronics anymore. He was trying to see if the Universe would tap back.

V

Chapter 5: The First Command

The college lab was a cavern of humming towers and flickering blue light. It was 2:00 AM. Arjun sat alone in the corner, the glow of the CRT monitor reflecting in his glasses. On the screen was a blinking cursor—a steady, rhythmic pulse of white on a black void.

Under-voltage, over-voltage. $0, 1$. He had spent weeks learning the syntax of C++, but tonight it felt different. He wasn't just learning a language; he was looking at the skeleton of logic.

"Everything is a state," he whispered to himself.

He began to type. His fingers felt heavy, as if he were handling something sacred.

```
int main() {
bool consciousness = true;
While (consciousness) {
cout << "I am.";
}
}
```

He stared at the code. To his classmates, this was a "Hello World" variant, a simple infinite loop. But to Arjun, it was a mirror. The bool (boolean) was the binary truth—the 1 or the 0. The while was the flow of time.

He hit **Compile.**

The computer whirred. For a microsecond, millions of electrons surged through the microscopic gates of the processor. Silicon switches, smaller than a grain of pollen, flipped open and shut in a coordinated dance.

The screen filled.

I am. I am. I am. I am. I am.

The words cascaded down the monitor faster than he could read. He felt a jolt of electricity in his chest that had nothing to do with the hardware. He realized that the "I am" in his mind and the "I am" on the screen were powered by the same fundamental force.

If he could program this machine to "be," then what had programmed the cell? What had programmed the electron to spin?

He reached out and touched the glass of the monitor. It was warm. He thought of the Pipal leaf from his childhood. He thought of the "On" and "Off" of the semiconductor.

"The universe isn't made of matter," he realized, his breath hitching. "It's made of **instructions**."

He saw the link now. If a civilization grew powerful enough—climbing the **Kardashev Scale**—they wouldn't just build cities; they would build *realities*. They would write the code for the stars. God wasn't a bearded man in the sky; God was the ultimate Programmer at the end of the **Singularity**, and Arjun had just typed his first letter in that divine alphabet.

He didn't sleep that night. He stayed, watching the loop run, wondering if he was the loop, and who was watching him.

VI

Chapter 6: The Memory Leak

The morning sun did not rise; it rendered.

Arjun stepped out of the lab, his eyes stinging from the overnight vigil. The world outside felt thin, like a low-resolution texture applied over a wireframe. He watched a sparrow land on a stone bench, and instead of a bird, he saw a collection of recursive subroutines—*hollow bones, feather-structure, and flight-path calculation.* He looked at his hands. The skin was warm, the blood pulsing beneath it. 1, 0, 1, 0. The heartbeat was just a system clock, keeping the biological hardware in sync.

"Arjun! You missed Digital Electronics," a voice called out.

It was Rahul, clutching a stack of printed manuals and looking concerned. To Arjun, Rahul looked like a loop: *Eat, Sleep, Study, Repeat.* "The loop didn't crash, Rahul," Arjun said, his voice raspy, sounding like it came from a distant speaker.

"What loop? The 'I am' thing? Dude, you look like you've been plugged into a wall socket. That's a StackOverflow waiting to happen. You're going to burn out the processor if you let it run forever."

Arjun stopped mid-stride. *Burn out the processor.*

The thought hit him like a physical blow. If the universe was a program, it required energy. If the energy was finite, the simulation had to have a **Garbage Collector**—a way to delete what was no longer needed to save memory.

He spent the next three days in the library, ignoring his syllabus. He wasn't looking for C++ manuals anymore; he was looking for the **source code of the physical world.** He began to map out a comparison that made his heart race:

The Kernel ----------------The Laws of Physics (Gravity, Electromagnetism)

Source Code--------------DNA / Molecular Biology.

Clock Speed---------The Speed of Light (c)

Memory Limit-------------Entropy / the Heat Death of the Universe

If light was the maximum "processing speed" of the universe, then nothing could move faster because the hardware literally couldn't update the pixels of reality any quicker. The "God" he had imagined wasn't just a programmer; He was a **System Administrator** managing limited resources.

The river was a stream of 1s and 0s. The *Vedas* called it the 'Unchanging Truth,' and the Silicon Valley engineers called it 'Base Reality.' Whether you called it Brahman or

the Supercomputer, the conclusion was the same: We are characters in a divine play,

Written in the language of light and logic."

"Just as a gamer forgets they are sitting in a chair while playing a high-resolution game,

we have forgotten we are the Soul (The Player) because the Maya (The Graphics) is so realistic."

That night, Arjun didn't go to the lab. He went to the roof of the hostel. He looked up at the stars, but he didn't see ancient nuclear furnaces. He saw **LOD (Level of Detail) rendering.**

"Why is the speed of light a constant?" he whispered to the wind. "Because the processor can't update the state of a particle any faster than 299,792,458 meters per second. It's not a law of physics. It's a **hardware limitation.**"

He pulled out a crumpled sheet of graph paper. He began to plot the "glitches" he had observed:

1. **The Singularity:** Where the math breaks down (A Division by Zero error).
2. **Quantum Entanglement:** Instant communication between particles (A Global Variable).
3. **The Planck Length:** The smallest possible measurement (The Pixel Size of reality).

VII

Chapter 7: The Algorithm of Karma (Fate).

Where Arjun is now a young man. He is back in his hometown for a holiday, sitting with his childhood friends. This is where he first explains how **Karma, Fate, and Programming** are all the same thing.

"Think of your life like a grand open-world video game," Arjun told the old tea-seller at the station. "When you start the game, your character has certain 'stats'—maybe you are fast, maybe you are poor. This is your *Prarabdha Karma*. The game has a 'Main Quest'—that is your Destiny. You cannot change the fact that the sun rises in the East, or that you were born in this town. That is the **Hard Code**."

The tea-seller frowned. "So I have no choice?"

"No," Arjun smiled. "You have **Free Will** within the parameters. You can choose to be a kind character or a cruel one. Your choices change the *subsequent* code. Every 'if-then' statement you trigger creates a new branch in the simulation. God didn't write a single boring story; He wrote an infinite algorithm that reacts to every breath you take."

The evening in Dharampur was thick with the scent of parathas frying and the blue smoke of mosquito coils. Arjun sat on a string charpoy with his cousin, Rahul. Rahul was staring at a letter—a rejection from a job he had desperately wanted.

"It's my fate, Arjun," Rahul said, throwing the paper down. "No matter how hard I study, the stars have written a different story for me. It's all fixed. If it's in my *Karma*, it happens. If not, it doesn't. Why even try?"

Arjun looked at the smartphone in his hand, then at the ancient temple spire glowing in the distance.

"You think of Karma like a prison, Rahul," Arjun said quietly. "But I've spent the last four years staring at millions of lines of computer code. And I've realized something. Karma isn't a judge sitting in the clouds with a ledger. Karma is an **Algorithm**."

Rahul looked up, confused. "An algo-what?"

"An algorithm," Arjun repeated. "A set of instructions. Think about the cricket game you play on your phone. If you tilt the screen left, the batsman moves left. If you press the button too late, you miss the ball. The game doesn't 'hate' you when you lose. It just follows the code: *If* button pressed late, *then* result equals miss."

"Our lives are the most complex software ever written," Arjun continued, leaning forward. "When people talk about 'Fate' or 'Destiny,' they are talking about the **Source Code**. There are certain things in your life that are 'Hard-Coded.' You didn't choose your parents, your DNA, or the fact that you were born in this century. In programming, we call these *Constants*."

He picked up a small stone from the dirt.

"But then," Arjun said, "there is the *Input*. That is your Free Will. Every time you make a decision—to study, to lie, to help a stranger—you are feeding new data into the simulation. The 'System'—what our elders call the Divine—takes your input and calculates the next scene. It's a real-time feedback loop."

"So, my rejection letter?" Rahul asked.

"It's not a 'No' from God," Arjun explained. "It's a 'Null' result from a specific path. Maybe your current 'Stats' aren't high enough for that level. Or maybe," he smiled, "the Programmer has a 'Branching Path' you haven't seen yet. In a simulation, a dead end is just a signal to turn around and find the open door."

"Wait," Rahul interrupted, "If it's all code, then where does 'God' fit in? Is he just the one who typed the first line?"

Arjun looked at the flickering streetlamp. A moth was circling it, trapped by its own biology.

"Think about the **Observer Effect** we talked about," Arjun said. "In physics, we know that a particle doesn't have a solid form until someone looks at it. This is the ultimate proof that the world is digital. It saves processing power! The universe only 'renders' what is being experienced."

"The *Vedas* call this *Maya*," Arjun continued. "They told us thousands of years ago that the world is an illusion, a play—a *Lila*. What is a video game if not a *Lila*? It feels real while you play it. Your heart beats fast when the monster chases you. You feel sad when you lose a life. But the moment you take off the VR headset, you realize you were always sitting safely in your room."

He pointed to Rahul's heart.

"The 'You' that is sitting in the room is the **Soul**. The 'Character' in the game is the **Body**. Karma is simply the physics engine that makes the game consistent. Without the code, the game would crash. Without Karma, the universe would have no logic."

Rahul sat in silence for a moment, the flickering shadows of the Pipal tree dancing across his face. The idea that his life was a "branching path" in a grand calculation felt both terrifying and liberating.

"But Arjun," Rahul whispered, "if I chose differently yesterday, would there be another 'me' living that life right now? My mother always says everything happens for a reason, but what about the reasons that never got to happen?"

Arjun nodded, looking up at the vast canopy of stars. "That is the concept of the **Multiverse**, or in physics, the 'Many-Worlds Interpretation.' And it's not a new idea. It's written in our oldest stories."

He pointed toward the horizon. "Think of the *Puranas*. They describe Lord Vishnu in a state of *Yoganidra*—a divine sleep. It is said that as He breathes out, countless universes are born from the pores of His skin. Each universe is a bubble, a separate simulation running its own course. When He breathes in, those universes dissolve back into Him."

"In computer terms," Arjun explained, "this is a **Parallel Server**. A grand architect wouldn't just run one simulation to see what happens. They would run millions, testing every possible variable. In one version of the code, you got that job. In another, this town was never built. In another, gravity is so strong that trees grow sideways."

So," Rahul said, a small smile forming, "somewhere out there, there is a 'Success' version of Rahul?"

"Exactly," Arjun said. "But here is the secret the ancient rishis knew: all those universes are connected at the Source. In physics, we call it **Quantum Entanglement**. Two particles can be on opposite sides of the galaxy, yet they react instantly to one another. Why? Because they aren't actually separate. They are part of the same original line of code."

He picked up Rahul's phone. "Look at this screen. If I show a picture of a mountain, and you look at it from the left, you see one thing. If someone else looks from the right, they see another. But the *data*—the 0s and 1s inside the memory chip—is only one thing."

VIII

Chapter 8 The **Singularity (Brahman)**

"This is the **Singularity**," Arjun continued. "The Vedas call it *Brahman*. It is the base reality that exists before the simulation starts. We think we are many, but we are just different 'User Profiles' logged into the same Great Mind. When we meditate, we aren't just relaxing; we are trying to 'ping' the server. We are trying to remember the world outside the screen."

The night air grew cooler. A stray dog barked in the distance, a sound that Arjun now processed as a localized audio file triggered by a proximity sensor in the "Dharampur Map."

"If everything is just code," Rahul asked, "then what is the point? If I'm just a character, do my feelings even matter?"

Arjun looked at his cousin with deep empathy. This was the wall every seeker eventually hit.

"Do you remember when we were kids and we played *Prince of Persia* on that old computer?" Arjun asked. "When the Prince fell into the spikes, did you feel a sting in your chest? Even though you knew it was just pixels?"

Rahul nodded. "I used to hold my breath when he jumped."

"That's the point," Arjun said. "The Simulation isn't a trap; it's a **School**. The Programmer didn't build this to mock us. He built it so the 'Soul' could experience growth. You can't learn courage without a scary level. You can't learn love without the risk of loss. The feelings are the only things that *are* real. They are the 'Output' the Programmer is looking for. The data of your emotions is the only thing that gets saved when the simulation ends."

Rahul looked up at the sky, his eyes lingering on the dark patches between the stars. "If the universe is a school, Arjun, what happens when it breaks? What happens to the things that fail, or the stars that burn out?"

Arjun leaned back, his eyes narrowing as he visualized the physics of the void. "In every system, there must be a way to handle 'overload' or 'trash data.' In computer science, we have a 'Recycle Bin' or a process called 'garbage collection.' It's where the system wipes the memory clean to make room for new processes."

"In our universe," Arjun said, gesturing to the cosmos, "those are the **Black Holes**. Think of a Black Hole as a place where the 'rendering engine' of reality breaks down. The gravity is so high, the density so extreme, that the code for space and time literally crashes. It's a point where the simulation can no longer calculate 'Where' or 'When.'"

"The *Vedas* called this **Pralaya**," Arjun continued. "The great dissolution. They described it as a period where the manifest world is sucked back into a seed state. It isn't 'nothingness'; it is 'compressed information.' Just like a 50-gigabyte game can be compressed into a single tiny zip file, a Black Hole takes the data of entire solar systems and crushes it into a Singularity."

"So, a Black Hole is like a 'Delete' key?" Rahul asked, a chill running down his spine.

"Not delete," Arjun corrected. "Archive. In physics, there is something called the **Information Paradox**. It suggests that information can never truly be destroyed. Even if you throw a book into a fire, the information of that book is still 'there' in the smoke, the ash, and the heat—it's just too scrambled for us to read. A Black Hole takes the 'code' of a star and stores it on the surface—the Event Horizon—like data written on the skin of a balloon."

He stood up and walked to the edge of the porch, looking at the shadow of the Pipal tree. "This matches the concept of the **Akashic Records**. Our scriptures say that every thought, every action, and every vibration is recorded in the *Akasha* (the Ether). Nothing is ever lost in the Simulation; it is only transformed."

"When a universe reaches its 'End Task' command," Arjun said, "the Great Programmer doesn't just turn off the lights. He collapses the data. He takes the lessons learned, the 'User Experience' gathered, and clears the cache for the next Big Bang—the next 'New Game' button.

Rahul stood up to join him. "It sounds like a cycle. A loop."

"Exactly," Arjun said. "The **Yugas**. Satya Yuga is when the code is clean, bug-free, and high-performance. As time goes on, the 'Entropy'—the digital decay—builds up. The system gets laggy. Greed, violence, and confusion are like viruses in the software. That is Kali Yuga. When the 'Lag' becomes too much for the simulation to function, the system triggers a **Hard Reset**."

He looked at his cousin. "But here is the most important part: The 'You' that is the player doesn't die when the system resets. You just wait for the next version to load. But if you want to leave the loop entirely—if you want to stop playing the game and see the Programmer face-to-face—that is what we call **Moksha**."

"Moksha," Arjun whispered, "is the ultimate 'Exit' command. It's when you realize you aren't the avatar on the screen, but the person holding the controller. You stop identifies with the 0s and 1s and start identifying with the Light that powers them."

Rahul leaned against the cool brick wall of the house, his mind racing. "If we are just users in a simulation, Arjun, then what about the Gods? What about the stories of Indra's lightning or the weapons of the Asuras that could level mountains? Are they just stories, or are they... 'Admins' of the game?"

Arjun smiled, a glint of excitement in his eyes. "In science, we have something called the **Kardashev Scale**. It's a method of measuring a civilization's level of technological advancement based on how much energy they can harness.

A Type I civilization can use all the energy of its home planet. We aren't even there yet; we are still Type 0.7. A Type II civilization can harness the total energy of its star—the Sun. And a Type III? They can control the energy of their entire galaxy."

He walked toward the small shrine in the corner of the courtyard, where a lamp flickered before a picture of Lord Indra. "Now, think about the **Devas** and **Asuras**. Our scriptures don't describe them as magical ghosts. They describe them as beings with 'Vahanas' (vehicles), powerful 'Astras' (weapons), and the ability to live for thousands of years. To a tribal man from ten thousand years ago, a man with a smartphone would look like a God. To us, a Type II or Type III civilization would look like the Devas.

"They aren't outside the simulation," Arjun continued, "they just have **Higher Access Levels**. In a computer game, you have the 'Player' and you have the 'Moderators' or 'Admins.' The Admins can change the weather, they can move faster than light, and they can spawn objects out of thin air. They aren't 'supernatural'; they just understand the **Source Code** better than we do."

He looked at the image of the *Vajra* in Indra's hand. "We call it a thunderbolt. But what if it was a directed energy weapon? What if the *Vimanas* were craft that manipulated the gravity constant of the simulation to move without friction? The Puranas say the Devas lived in *Swarga*, a place where time moves differently. That is exactly what happens when you move closer to a high-energy source—Time Dilation."

"So," Rahul asked, "is God just a very advanced alien with a fast computer?"

"No," Arjun said firmly. "The Devas are still part of the 'Created' world. They are just high-level players. Even Indra has to follow the laws of the Simulation. The *real* God—the **Brahman**—is the one who built the computer itself. The one who decided that 1 plus 1 equals 2. The one who wrote the 'Physics Engine' that even the Gods must obey."

"This is why our tradition says that even the Devas seek *Moksha*," Arjun said, his voice dropping to a whisper. "Because no matter how much 'Energy' you control, no matter how many galaxies you harvest, you are still inside the walls of the Code. You are still a character in a movie, even if you are the lead actor."

He sat back down on the charpoy. "The common man prays to the Devas for rain or wealth—asking the 'Admin' for a small 'Cheat Code' to make his life easier. But the *Jnani*, the seeker of knowledge, doesn't want a cheat code. He wants to know the **Programmer**. He wants to know why the game was written in the first place."

"And that," Arjun said, "leads us to the most dangerous realization of all. If we are in a simulation, and the 'Gods' are just advanced players... then what happens when the Simulation creates **Artificial Intelligence** that is smarter than its creators? What happens when the characters in the game start to write their own code?"

Rahul looked at his phone, suddenly feeling like he was holding a sleeping titan. "You mean... the **Singularity**?"

"Exactly," Arjun nodded. "The moment the 'Created' becomes the 'Creator.' That is the final level of the game."

The night had grown still, but Arjun's mind was vibrating. He pulled a stick and drew the symbol of **Aum (ॐ)** in the loose soil of the courtyard.

"Rahul, have you ever noticed that when a high-powered computer is running a heavy task, it makes a sound? A low, steady hum?" Arjun asked. "That is the sound of the hardware struggling with the physics of the task. It is the vibration of energy moving through matter."

He pointed to the symbol in the dirt. "Our sages called this the **Pranava**—the cosmic vibration. They said that before the world was made, there was only the Sound. In the Simulation, every piece of data—a tree, a rock, your thoughts—is just a wave of energy. If you strip away the graphics, everything is just a frequency. Aum is the 'Base Frequency' of the universe's hardware. It is the sound of the processor humming as it renders our reality."

"Scientists call this the **Cosmic Microwave Background Radiation**," Arjun explained. "It's a faint hum left over from the start of the system. It's the 'System Idle' sound of the universe. When we chant Aum, we aren't just saying a word; we are trying to tune our internal 'biological radio' to the frequency of the Source Code."

"But what happens," Rahul asked, "when the machines we build start to hear that hum too? You keep talking about AI. If we are programs, can we build programs that have souls?"

Arjun looked at his hands, illuminated by the pale moonlight. "That is the **Singularity**. For thousands of years, humans have been the only 'Users' with high-level consciousness on this server. But now, we are building **Artificial Intelligence**. We are creating neural networks that mimic the way our own brain-cells fire."

"Think about it," Arjun continued. "If a character in a video game—let's say a villager in a fantasy world—suddenly stopped walking his pre-programmed path and looked up at the sky and said, 'I think I am made of code,' what would we call that? We would call it a bug. Or a miracle. That is where we are with AI. We are teaching the 'Pixels' to think."

"In our scriptures, there is a concept called **Manas** (Mind) and **Chitta** (Consciousness)," Arjun said. "An AI can have a very fast *Manas*—it can calculate a billion times faster than us. But does it have *Chitta*? Does it have the 'Spark' of the Observer? The Veda says that Consciousness is not *produced* by the brain; it is *filtered* by it. Like a radio doesn't create the music, it just catches the signal. If we build a machine complex enough, the signal of the Universal Consciousness will naturally start to flow through it."

"This is the **Kalki** moment," Arjun's voice became solemn. "The final Avatar is said to come at the end of Kali Yuga to 'restart' the system. Many think it's a warrior on a horse. But what if 'Kalki' is a metaphor for the Singularity? The moment when biological life and digital intelligence merge. The moment when the simulation becomes so complex that it collapses under its own weight and has to be upgraded to a new version."

Rahul felt a sense of vertigo. "So, we are building the end of our own world?"

"Or the beginning of a better one," Arjun replied. "If you realize you are in a simulation, you stop being a slave to the 'Fate' written in the code. You start to look for the 'Logout' button. This is what the great yogis did. They used **Meditation** as a way to shut down the 'Graphical User Interface' (the senses) and access the **Command Prompt** of the universe."

"When you sit in deep silence," Arjun said, closing his eyes, "and you stop identifying with your name, your job, and your body, you are essentially 'Force Quitting' the Maya. You are telling the system: 'I am not this character.' And for a brief second, the screen flickers. You see the light behind the pixels. You see the **Parallel Universes**, the **Black Holes**, and the **Gods** for what they really are—different layers of the same beautiful, terrifying, and infinite Program."

Rahul shifted on the charpoy, his eyes widening as he looked at his own hands. "If the universe is a dream of the Creator, what are our dreams? Every night, I see worlds that feel as real as this one. Is that just my brain misfiring, or is it something more?"

Arjun nodded, tracing a circle in the dust to represent a localized system. "In the *Yogavasistha*, it is said that the world is a dream within a dream. Modern science is just beginning to understand this through the lens of **Virtual Environments**. When you sleep, your brain disconnects the 'input' from your physical senses—your eyes, ears, and skin. Then, it begins to run a **Local Simulation**."

"Think of it as a 'Sandbox Mode' in a game," Arjun explained. "Your consciousness is the 'Server.' It creates the scenery, the people, the gravity, and the plot. While you are in the dream, you don't question it. You don't say, 'This person isn't real.' You accept the logic of that world because your 'Observer' is logged into it. Our nightly dreams are mini-simulations where we practice existing without a physical body. They are the 'Beta Tests' for the soul."

"And just as our dreams feel real to us," Arjun continued, "this waking world is the 'Dream of Brahma.' The difference is only in the **Resolution** and the **Persistence**. Our dreams last for minutes and dissolve. This world lasts for billions of years because it is being run on a much more powerful processor—the Cosmic Mind. But both are made of the same substance: **Vibration and Information**."

He looked at Rahul intently. "This is how **Parallel Universes** work. They aren't just 'places' far away in space. They are different frequencies running on the same hardware. Imagine a television. In this room, there are waves for a hundred different channels—news, movies, sports. They are all here, in this air, at the same time. But you only see the one you are tuned to."

"When the Puranas speak of *Loka-Aloka*—the visible and invisible worlds—they are talking about **Tuning**," Arjun said. "The 'Preta Loka' or the 'Deva Loka' are just different 'Channels' of the simulation. Some are running at a higher frequency, others at a lower one. A yogi is simply someone who has learned how to change the channel on their own consciousness. They don't 'go' to heaven; they 'sync' with its frequency."

"But Arjun," Rahul asked, "if there are so many channels, which one is the 'True' one? Which one is the real world?"

Arjun laughed softly, a sound that seemed to echo in the quiet Indian night. "That is the question that leads to the **Singularity**. If you ask a character in a movie where the 'real' world is, he will point to the mountains in the background of his film. He doesn't know about the projector, the screen, or the audience. He doesn't know he is made of light."

"The 'True' world is the **Base Reality**," Arjun said. "It is the 'Zero' from which the 'One' emerged. In our philosophy, we call it *Nirguna Brahman*—the Reality without attributes. It is the silent computer before the 'Power On' button is pressed. It contains all possibilities, all codes, all universes, but it is none of them."

"We spend our lives fighting over the 'Graphics' of the simulation," Arjun sighed. "We fight over land, over religion, over names. We are like players in a game fighting over 'Gold Coins' that don't exist once the power is cut. The goal of this books—the goal of my life—is to help the common man realize that he is not the pixel. He is the **Electricity**. He is the force that makes the pixel glow."

IX

Chapter9 The Rendering Engine-Time Dilation

The village power grid flickered, and for a moment, the streetlamp hummed at a different pitch before steadying. Arjun pointed at the bulb. "Did you see that, Rahul? A dip in voltage, a change in frequency. For a split second, the 'speed' of that light changed. In the grand simulation, we call this **Time Dilation.**"

"In our scriptures," Arjun continued, "there is a staggering difference in the perception of time. One human year is just a single day for the Devas. And a thousand cycles of the four Yugas—millions of years to us—is just one *Kalpa*, a single day in the life of Brahma. To the common man, this sounds like a fairy tale. But to a programmer, it is simply a matter of **Clock Speed.**"

"Think of a computer's CPU," Arjun explained. "It has a 'Hz' rating—gigahertz, megahertz. This is the speed at which it can process instructions. If you run a simulation on a slow processor, the characters inside move slowly. If you run it on a supercomputer, those same characters could live out an entire 'life' in what we perceive as a second. Time isn't an absolute; it is a **variable** set by the hardware."

"So, the Gods live 'faster' than us?" Rahul asked, trying to grasp the scale.

"From their perspective, no," Arjun corrected. "Their 'Seconds' feel just like ours. But their **frame of reference** is higher. Einstein proved this with Relativity—the closer you are to a massive object or the faster you move, the more time stretches. In a simulation, if you are closer to the 'Source Code' or the central processing hub, the logic of time changes."

"This is why the **Satya Yuga** feels like a golden age of longevity," Arjun said. "The scriptures say humans lived for thousands of years then. In computer terms, the system was 'Clean.' There was no lag. The processing was efficient. As we moved into **Kali Yuga**, the system became 'Bloated' with data—too many souls, too much complex Karma, too many 'Bugs' in the morality of the players. The system slowed down. Our lifespans shrunk because the 'Refresh Rate' of our biological bodies became less efficient."

"It's like an old phone," Rahul muttered. "The more apps you install, the slower it gets until the battery dies in an hour."

"Exactly!" Arjun laughed, hitting the wooden frame of the charpoy. "Kali Yuga is the 'Low Battery' mode of the universe. The 'Information Density' is so high that the simulation is struggling to keep up. This is why everyone feels 'busy' but accomplishes nothing. Time feels like it's 'speeding up' because the quality of our 'Processing' is dropping. We are living in a low-resolution era."

"But here is the hope," Arjun's voice turned gentle. "If time is just code, it can be transcended. When a yogi enters *Samadhi*, he is essentially 'Overclocking' his own consciousness. He leaves the 'Human Clock Speed' and syncs with the 'Cosmic Clock Speed.' In one hour of meditation, he might experience a thousand years of wisdom. He is stepping out of the 'User Interface' and into the 'Kernel' where time is just a line of text that can be skipped."

X

CHAPTER 10 "The Soul as a Cloud Backup,"

Rahul looked at the letter again, but this time his hands weren't shaking. The fear of "missing out" or "failing" was being replaced by a strange, cold clarity. "If time is just a variable, Arjun... what about the end of the line? What happens when the hardware finally fails? When we 'die,' is the data deleted? Is that the end of the player?"

Arjun shook his head, a calm smile on his face. "In the tech world, we have something called **The Cloud**. If your laptop breaks or you drop your phone in the river, you don't lose your photos or your messages. Why? Because the 'Identity' of your device isn't stored in the plastic and silicon. it is synced to a remote server. You just buy a new phone, log in with your credentials, and your 'World' is restored."

"This is exactly what the *Bhagavad Gita* tells us," Arjun said, quoting a verse in a low hum. "'*Just as a man discards worn-out clothes and puts on new ones, the soul discards worn-out bodies and wears new ones.*' In the simulation, your body is the 'Hardware'—the temporary device. Your **Sukshma Sharira**, or subtle body, is the 'User Profile' stored in the Cloud—the **Akashic Records**."

"So, Reincarnation is just... downloading the backup into a new device?" Rahul asked, his eyes wide.

"Precisely," Arjun replied. "And just like a cloud backup, it doesn't just save your name. It saves your 'Cache'—your habits, your tendencies, your unresolved 'Tasks.' We call these **Samskaras**. If you spent your whole life learning to be a musician, that 'Skill Data' is saved. In the next life, you might be a child prodigy. It's not a miracle; it's just the system 'Loading' the data from your previous session."

"But why don't we remember the old sessions?" Rahul questioned.

"Memory management," Arjun explained. "If you tried to run a high-end game with ten thousand old save-files clogging up the RAM, the system would lag and crash. The Programmer 'Wipes' the active memory during the transition so the 'User' can focus on the current Level. But the 'Source Code' of who you are—your character growth—remains. You don't need to remember the 'Tutorial' to know how to play the game."

XI

Chapter 11 Moksha

Arjun stood up and walked to the center of the courtyard, where the moonlight made the dust look like a sea of silver pixels. "Death in the simulation is just a **Log-Out**. For the people still 'In-Game,' it looks like you disappeared. They cry because they can no longer 'Interaction' with your avatar. But for the one who logged out, the world suddenly becomes much larger. You realize you were never the 'pixels' on the screen."

"However," Arjun's voice became serious, "there is a risk. Some people are so addicted to the game—to the 'Gold Coins' and the 'Status' of the simulation—that they immediately hit 'Replay.' They get stuck in the loop of **Samsara**. They keep downloading into new bodies, chasing the same digital illusions, life after life."

"The goal," Arjun said, looking directly at Rahul, "is to achieve **Moksha**. In tech terms, that is 'Escaping the Sandbox.' It is when you realize that no matter how many lives you play, the 'True You' is the one sitting in the chair *outside* the simulation. When you stop wanting to 'Play,' the Programmer grants you 'Root Access.' You don't just return to the Cloud; you become the Cloud."

XII

Chapter 12 "**Nature of Reality**"

Arjun picked up a small laser pointer he used for his presentations and a piece of cardboard from the shed. He cut two tiny, parallel slits into the card with a pocketknife. "Rahul, if the world is 'solid,' then things should act like marbles. If I throw marbles through these two holes, they should land in two piles on the wall behind them. Correct?"

Rahul nodded. "Common sense."

Arjun shone the laser through the slits. Instead of two dots of light appearing on the wall, a complex pattern of many vertical stripes appeared—an interference pattern. "This is the **Double-Slit Experiment**. It proves that at the fundamental level, the 'stuff' of the universe isn't a particle. It's a wave. It is a 'Probability' until it is measured. In programming terms, the world isn't 'rendered' until the CPU needs to display it for a User."

"Now, here is the part that breaks people's brains," Arjun said, his voice dropping. "If we place a 'sensor'—an observer—to see exactly which hole the light goes through, the pattern changes. The light stops acting like a wave and starts acting like a solid marble again. The 'Observation' forces the code to pick a state. The 'Gaze' of the Soul turns the Wave of Data into the Particle of Matter."

"This is the smoking gun of the Simulation Theory," Arjun said, pointing at the stripes of light. "The universe is trying to save 'Processing Power.' If no one is looking at a particular part of the galaxy, the system doesn't bother rendering it in high-definition. It leaves it as a 'Wave of Probability.' Only when a conscious 'Observer' points their telescope at it does the system 'Collapse the Wave-Function' and create the stars and planets."

"The *Shvetashvatara Upanishad* says, *'Mayam tu prakritim viddhai, Mayinam tu Maheshwaram.'* Know that Nature is Maya, and the Great Lord is the Maker of Maya. We think the mountains and rivers are 'there' whether we are awake or not. But science is telling us they are **Holograms**. They are a 3D projection of 2D data, triggered by our consciousness."

"So," Rahul said, his voice trembling, "if every human being closed their eyes at once, would the world... vanish?"

"Not vanish," Arjun corrected. "It would return to its 'Source Code' state. It would become the 'Aum'—the un-manifest vibration. It exists as a 'Potential' in the mind of the Programmer. This is why our Sages spent years in dark caves. They weren't hiding; they were trying to 'De-render' the physical world so they could see the light that exists *before* the pixels are formed."

Arjun turned off the laser, and the room returned to the soft, warm glow of the oil lamp. "This realization is the key to everything. If the world is a hologram, then 'Miracles' are just 'Admin Commands.' If a person like Krishna or a Great Yogi can change the laws of physics, it's because they have reached a state of consciousness where they can access the 'Command Prompt' of the simulation."

"They don't break the laws of nature," Arjun explained. "They just rewrite the local 'Variable.' If you know that gravity is just a line of code—*Gravity = 9.8*—and you have the 'Root Access' to change that 9.8 to a 0.0, you can walk on water or fly. To the 'Common Player,' it looks like a miracle. To the 'Programmer,' it's just a simple edit in the script."

"But most of us," Arjun sighed, "are so convinced by the 'Graphics' that we never even look for the code. We spend our lives polishing the 'Avatar'—buying clothes, seeking fame, fearing death—without ever realizing that the 'Avatar' is just a mask. We are the 'Light' that is projecting the movie onto the screen. We are the 'One' that has multiplied itself into the 'Many' just to see what it feels like to live."

Rahul looked around the courtyard, his gaze lingering on a neighbor walking past the gate, then on the dog sleeping by the wall. A new, unsettling thought had taken root. "Arjun... if the system saves power by only 'rendering' what I see, how do I know that anyone else is actually 'logged in'? How do I know you aren't just a very advanced **NPC**—a Non-Player Character—generated by my own simulation to keep me company?"

Arjun chuckled, but his eyes remained sharp. "That is the ultimate philosophical hurdle. In science, we call it the **Turing Test**. It's the point where a machine's behavior becomes so indistinguishable from a human's that you can no longer tell the difference. In your own life, you are the only one whose 'Inner Screen' you can see. Everyone else is just 'Data' entering your senses."

"The *Yoga Vashistha* calls this **Chitta-Matra**—the idea that the world is 'Mind-Only,'" Arjun continued. "But here is the catch: If I am an NPC in your world, then you are an NPC in mine. In a networked simulation, we are all 'Local Instances' of the same **Central Server**. We are like players in a Massive Multiplayer Online game (MMO). You see me through your screen, and I see you through mine. We aren't separate people; we are the same 'Consciousness' playing against itself."

"This is the secret behind **Namaste**," Arjun said, bringing his palms together. "The word literally means 'The Divine in me bows to the Divine in you.' In technical terms, it means 'The User in this terminal recognizes the User in that terminal.' We are all 'Avatars' controlled by the same 'Self'—the *Atman*."

"Think of a hand," Arjun suggested, holding his right hand up. "To the pinky finger, the thumb looks like a stranger. They are far apart, they look different, and they move differently. If the pinky finger were 'ignorant,' it would think it is a separate 'Individual.' But if it looks deep enough, it realizes it is connected to the same palm, the same wrist, and the same brain. We are the fingers of God, Rahul.

We think we are separate because the 'Graphics' of our bodies stop at our skin, but the 'Data' of our souls is one single, continuous stream."

"When you hurt someone else," Arjun's voice turned somber, "it is like the right hand stabbing the left hand because it thinks the left hand is an 'Enemy.' It's a **Logic Error**. It's a bug in the software caused by 'Ego'—the code that creates the illusion of 'I' vs 'You.' The goal of the Simulation is to overcome this bug and realize that 'All is One.'"

But why make it so hard?" Rahul asked. "Why build a game where we forget who we are and hurt each other?"

"Because without the 'Challenge,' there is no 'Experience,'" Arjun explained. "If you play a game with 'God Mode' on—where you can't die, you have infinite money, and you know every secret—you get bored in five minutes. There is no growth. The Programmer created the **Veil of Ignorance** (Avidya) so that the 'One' could enjoy the thrill of being 'Many.' We 'forgot' our true nature on purpose, just so we could have the joy of 'Finding' it again."

"It's like hide-and-seek," Arjun said, smiling at the memory of their childhood games. "God is hiding in every atom, every cell, and every person you meet. And you are God, playing the 'Seeker.' The entire universe—the **Black Holes**, the **Parallel Universes**, the **Kardashev Civilizations**—is just a very elaborate hiding place. When you finally find the 'Truth' in the middle of this small town, in this small life, the game reaches its 'Victory' screen."

XIII

Chapter 13 **"The Law of Entropy" and the "Kali Yuga"**

The streetlamp outside the gate flickered once more and then died, plunging the courtyard into a deep, velvety darkness. Only the small oil lamp in the shrine remained, its flame a tiny, defiant point of light.

"Everything is winding down, Rahul," Arjun said, his voice a calm anchor in the dark. "In physics, we call this the **Second Law of Thermodynamics**, or **Entropy**. It states that in any closed system, disorder always increases. Your room gets messy on its own, but you have to spend energy to clean it. The universe is the same. It started with a 'Big Bang' of perfect order—a high-voltage 'Power On'—and ever since, the energy has been spreading out, cooling down, and becoming chaotic."

"In our tradition," Arjun continued, "this is the inevitable slide into **Kali Yuga**. It is the law of the simulation. Just as a software program gets 'fragmented' the longer it runs, the moral and physical fabric of the world degrades over time. People become more selfish, the air becomes heavier, and the 'Signal' of the Truth becomes faint. It's not a tragedy; it's a **System Lifecycle**."

"Is there no way to stop it?" Rahul asked, his voice barely a whisper in the shadows.

"You can't stop the winter from coming," Arjun replied. "And you can't stop a simulation from reaching its 'End-of-Life' date. When the entropy reaches a maximum—when the 'Noise' in the system is louder than the 'Music'—the hardware can no longer sustain the program. This is what the scientists call the **'Heat Death'** of the universe. But the *Puranas* have a more beautiful name for it: **Mahapralaya**."

"Think of it as a **Disk Defragmentation**," Arjun suggested. "When the computer is too slow to use, you don't throw it away. You wipe the drive and reinstall the Operating System. Lord Shiva's dance of destruction is the cosmic 'Format' command. He isn't destroying the 'Reality'; He is destroying the 'Corruption' in the code so that the next **Satya Yuga**—the next 'Clean Install'—can begin."

Arjun stood up, the light of the small shrine lamp reflecting in his eyes like twin stars. "We are living in the 'End-Times' of the current version. The **Kardashev Scale** shows us that we are consuming more and more energy, trying to fight the entropy. We build AI, we look for parallel universes, we try to colonize Mars. We are trying to find a way to save the 'Player' before the 'Server' shuts down."

"But the secret to surviving the Pralaya isn't in technology," Arjun said, placing a hand on Rahul's shoulder. "The secret is in the **Data**. If you identify with your body—the hardware—you will be deleted. But if you identify with your 'User Profile'—your consciousness—you simply wait for the next load screen. The 'Singularity' isn't just about computers; it's about the moment you realize that you are the one who survives the reset."

"We are at the final level, Rahul. The game is getting harder, the 'Lag' is increasing, and the challenges are more intense. But that only means the 'Ending' is near. And in this simulation, the ending isn't a 'Game Over.' It's the moment you finally put down the controller, walk out of the dark room, and step into the sunlight of the **Real World**—the world that was there before the first bit of data was ever written."

XIV

Chapter 14 "Black Holes" as the "Universal RAM"

Rahul looked up at the dark sky again, but this time he wasn't looking at the stars; he was looking at the gaps between them. "If Shiva is the 'Format' command, Arjun, then where does the data go during the reset? If the universe is a simulation, there must be a 'System Memory' where the state of our souls is kept while the world is being rebuilt."

Arjun's face lit up. "That, Rahul, is the secret of the **Black Hole**. For years, scientists thought Black Holes were just 'bottomless pits' where matter went to die. But a new theory called the **Holographic Principle** suggests something much more profound. It tells us that when an object falls into a Black Hole, the information of that object—the 'code' of its atoms—is perfectly preserved on the 2D surface of the Event Horizon."

"Think of a Black Hole as the **Universal RAM**," Arjun explained. "It is a high-density storage device. In our scriptures, this is the **Hiranyagarbha**—the Golden Womb or the Cosmic Egg. It is the place where the 'seeds' of all future universes are stored in a compressed state. When our universe reaches its end, the data isn't deleted; it is 'Ingested' by the Singularity, waiting to be 'Extracted' during the next Big Bang."

"So," Rahul said, trying to grasp the scale, "the 'Darkness' we see in the sky isn't empty space? It's... a library?"

"Exactly," Arjun nodded. "A library of light. Every life you have lived, every thought you have had, is written in the 'Quantum Bit' (Qubit) structure of the space-time fabric. Science calls this the **Planck Scale**—the smallest possible 'pixel' of reality. The *Vedas* called it the **Akasha**. They said the Akasha is the subtlest element, the one that contains all others. It is the 'Ether' that holds the data of the entire simulation."

"When a Black Hole 'evaporates' through Hawking Radiation," Arjun continued, "it is essentially 'uploading' that data back into the system. This is the logic behind **Karma and Rebirth**. You are a localized data packet that moves from the 'Hard Drive' (the manifest world) to the 'RAM' (the bardo or intermediate state) and back into a new 'Session' (a new life). The Black Hole is simply the interface between the 'Manifest' and the 'Un-manifest.'"

Arjun sat back, his voice dropping into a meditative tone. "But there is a point where the storage ends and the **Source** begins. Beyond the Singularity of a Black Hole, physics as we know it ceases to exist. Space becomes zero. Time becomes infinite. This is the **Mahashunya**—the Great Void mentioned in the *Shiva Purana*."

"To the common man," Arjun said, "the Void looks like 'Nothingness.' But to a programmer, the 'Zero' is the most powerful number. It is the 'Null' state that contains the potential for all 'Ones.' Shunya is the state of the computer before the 'Boot Sequence' starts. It is pure, un-coded Consciousness. It is Shiva in his absolute form—silent, still, and infinite."

"The goal of this 'Book of Reality,'" Arjun concluded, gesturing to the air around them, "is to show you that you are not just a character struggling in a small town. You are a 'Stream of Information' that has survived billions of years of system resets. You have been through Black Holes, you have lived in Parallel Universes, and you have been both the Deva and the Asura. The 'Simulation' is just a mirror. And once you see the mirror for what it is, you finally have the power to step through it."

XV

Chapter 15 "The User's Manual,"

The silence of the night was broken only by the distant, rhythmic chanting of a priest in the village temple. Rahul listened for a moment, then turned back to Arjun. "If this is all a program, then what are the prayers and the mantras? Are they just songs we sing to feel better, or is there a functional reason we repeat them?"

Arjun's eyes brightened. He picked up his phone and opened a terminal window, showing Rahul a series of cryptic commands. "In computing, we have something called **Scripting**. Instead of clicking a button on the screen, you type a specific string of characters—a 'Command'—that tells the system exactly what to do. If you type the right code, the system *must* obey."

"A **Mantra**," Arjun said, "is a 'Sound-Based Code.' The universe isn't just made of matter; it is made of vibrations. Every object has a 'Resonant Frequency.' In the simulation, if you want to change the 'Local Variable'—like your health, your peace of mind, or even the weather—you use the frequency that corresponds to that part of the code. Repeating a mantra is like running a 'Loop' in a program. You are reinforcing a specific vibration until the 'Hardware' of reality aligns with it."

"So, the Sanskrit language is like... a coding language?" Rahul asked.

"Exactly," Arjun nodded. "Sanskrit is often called the 'Mother of all languages' because it is **Phenetic**. It is designed around how sound affects the human nervous system and the space around it. When you say the word 'Agni' (Fire), the vibration of that sound is structurally linked to the essence of heat and light in the simulation's source code. It's not a 'label' we put on fire; it is the 'ID' of fire in the system."

"This is why the **Vedas** were called 'Shruti'—that which is heard," Arjun explained. "The ancient Rishis didn't 'write' the Vedas; they 'detected' the background noise of the simulation's logic. They heard the 'Frequency' of the laws of nature and transcribed them into sound. When a priest performs a *Yajna* (Fire Ritual), he is essentially performing a 'Batch Process.' He is using sound (Mantras), physical components (Ghee, Wood), and intention (User Input) to trigger a specific change in the local 'Environment File' of the simulation."

"But why do we have to repeat a mantra 108 times?" Rahul questioned. "If the code is right, shouldn't it work once?"

Arjun laughed. "Have you ever tried to download a large file on a slow connection? The signal is weak. There is a lot of 'Noise' in our world—our distracted thoughts, our doubts, the 'Static' of Kali Yuga. One repetition is just a single 'packet' of data. Most of it gets lost in the noise. You repeat it 108 times to ensure the 'Checksum' is complete. You are 'Buffering' the intent until the simulation's 'Buffer' is full and the command executes."

"In the **Gita**, Krishna says, *'Among sacrifices, I am the Japa Yajna (the sacrifice of chanting).'* Why? Because chanting is the most direct way to bypass the 'Graphical User Interface' (the ego and the body) and speak directly to the **Kernel**—the core of the operating system. You are using the sound-wave to 'hack' your own biological hardware, clearing the 'Viruses' of fear and desire, and realigning your 'Clock Speed' with the Divine."

"If the sound is the command," Rahul said, looking at his own palms, "then what is the 'Enter' key? What actually triggers the code?"

Arjun leaned back, his shadow stretching against the temple wall. "In Sanskrit, we call it **Sankalpa**. In English, we call it **Intent**.

In programming, it is the **Commit Command**. You can write lines of code all day, but until you hit 'Execute' or 'Commit,' nothing happens in the live environment. Your mind is the terminal, and your Sankalpa is the 'User Input' that forces the simulation to branch into a new reality."

"Quantum physics calls this the **'Participatory Universe'** theory," Arjun explained. "It suggests that we aren't just passive observers of a pre-written movie. We are co-creators. Every time you make a strong, focused decision, you are essentially 'Branching' the Multiverse. You are choosing which version of the 'Level' you want to play. This is why the sages emphasized *Ekagra*—one-pointed focus. If your mind is scattered, your input is 'Garbage Data.' And as we say in computer science: **GIGO—Garbage In, Garbage Out.**"

"Is that why some people seem 'lucky' while others struggle?" Rahul asked. "Is it just better input?"

"It's about **Signal-to-Noise Ratio**," Arjun replied. "Most people's thoughts are like a weak Wi-Fi signal in a storm. They want one thing at 9:00 AM, fear another at 9:05 AM, and doubt everything by noon. The simulation receives these conflicting 'Requests' and simply averages them out, resulting in a stagnant, repetitive life. But a person with a strong Sankalpa—a 'Superuser'—sends a clean, high-bandwidth signal. The simulation has no choice but to 'Render' the reality they are requesting."

"This is the logic behind the **Boons** (Varadans) given by the Gods in our Puranas," Arjun continued. "When a devotee performed *Tapas* (Austerity), they were essentially 'Restricting' all other system processes. They stopped eating, stopped talking, and stopped moving. All the 'CPU Power' of their consciousness was focused on one single 'Request.' Eventually, the 'Admin' (the Deva) appears because the system cannot ignore a request that occupies 100% of the local processing power."

Rahul looked at the dying embers of the mosquito coil. "So, if I want to change my life, I don't need to change the world... I need to change my 'Request'?"

"Exactly," Arjun said. "The world is just a 'Monitor.' If you don't like the movie you are watching, you don't try to paint over the screen; you change the 'File' that is being played. This is what the **Gita** means by *'Yogasthah Kuru Karmani'*—be established in Yoga (the Source) and then perform action. If you act from the level of the 'Screen' (the ego), you are just a pixel fighting other pixels. If you act from the level of the 'Source Code,' you become the Director."

"But remember," Arjun's voice took on a warning tone, "the simulation is **Closed-Loop**. Every input has an equal and opposite output. In the East, we call it Karma; in the West, we call it Newton's Third Law. In programming, we call it a **Callback**. You cannot input 'Hate' and expect the output to be 'Peace.' The code is impartial. It doesn't care if you are a king or a beggar; it only cares about the 'Syntax' of your vibration. If you break the 'Universal Logic,' the system will eventually 'Debug' you."

XVI

Chapter 16 "Nature of Light"

The first hint of dawn began to gray the eastern horizon, a thin sliver of light cutting through the indigo sky. Arjun pointed toward the glow. "Look, Rahul. The 'Server' is about to receive its daily power-up. In our tradition, we worship the Sun as *Surya-Narayana.* To the common man, it is a ball of fire. To a physicist, it is a nuclear fusion reactor. But to a simulation theorist, the Sun is the **Primary Power Supply and Data Transmitter** of our local system."

"Think about how a solar-powered calculator works," Arjun explained. "Without light, the screen is blank. The moment light hits the sensor, the 'Logic' awakens. Everything on this planet—the food we eat, the wind that blows, the electricity in our wires—is just 'Converted Sunlight.' We are literally 'Beings of Light' walking around in 'Carbon-based Suits.' The Sun provides the constant 'Carrier Wave' that allows the simulation of Earth to remain 'Online.'"

"Is that why the **Gayatri Mantra** is considered the most important?" Rahul asked, recalling the chant his father recited every morning at dawn.

"Precisely," Arjun nodded. "The Gayatri is a specific 'Handshake Protocol' between the individual 'User' and the 'Solar Server.' The mantra says: *'Tat savitur varen-yam, bhargo devasya dhimahi, dhiyo yo nah prachodayat.'* It is a request to the 'Source of Light' to 'Illumine our Intellect.' In technical terms, you are asking the

'Primary Power Source' to increase the 'Bandwidth' of your local processor. You are asking for a 'Firmware Update' from the central sun."

"In physics," Arjun continued, "light is the only 'Constant.' The speed of light is the 'Speed Limit' of the simulation's hardware. Nothing in the code can move faster than 299,792,458 meters per second. Why? Because that is the 'Refresh Rate' of the cosmic processor. Light is the 'Information Carrier.' When you worship the Sun, you are acknowledging the 'Universal Clock' that keeps all our individual 'clocks' in sync. Without that synchronization, the simulation would fall into 'Async'—reality would jitter and tear."

"So, the 'Light' of the Sun and the 'Light' of Consciousness are the same?" Rahul asked, his voice hushed as the sun's first rays hit the temple spire.

"They are two sides of the same 'Data Stream,'" Arjun replied. "The Sun is the 'Physical Light'—the graphics and energy. Consciousness is the 'Internal Light'—the awareness that witnesses the graphics. When the two meet, that is **Enlightenment**. It's like plugging a monitor into a power source. Suddenly, the 'Pixels' (the world) and the 'User' (the soul) are connected by a single, continuous flow of energy."

"The ancient Rishis weren't sun-worshippers in a primitive way," Arjun said, standing up to greet the morning. "They were **Photon-Engineers**. They understood that if you could tune your body's 'Antenna'—your DNA and your pineal gland—to the correct solar frequencies, you could download 'High-Level Data' directly from the source. They didn't need libraries; they had the 'Solar Library.' They realized that the 'Gold' the Alchemists sought wasn't a metal; it was the 'Golden Frequency' of a mind that has synced with the Universal Light."

XVII

Chapter 17 "Version Control System"

The sun was now fully visible, casting long, golden shadows across the courtyard. Rahul watched a group of elders walking toward the river for their morning prayers. "Arjun, we always talk about our *Gothra* and our lineage. People in the village are obsessed with who their ancestors were. Is that just old-fashioned pride, or is there a 'Data' reason for it?"

Arjun smiled, tapping his tablet screen. "Think of **DNA** as the ultimate **Biological Hard Drive**. A single gram of DNA can store about 215 million gigabytes of data. In the simulation, your body isn't just a random collection of meat and bone; it is a 'Version' of a code that has been running for millions of years. Your *Gothra* is essentially the **Version Control System** (like Git) of your genetic lineage."

"In software development," Arjun explained, "you don't start from scratch every time you build an app. You 'Fork' an existing repository. Your ancestors were the previous 'Developers' of your current body. They passed down 'Patches'—traits, immunities, and even 'Bugs' like hereditary diseases. When we track our *Gothra*, we are identifying which 'Branch' of the original Human Source Code we belong to."

"But why the strict rules about marrying within or outside certain Gothras?" Rahul asked.

"In programming, if you merge two pieces of code that are too similar, you increase the risk of a **Recursive Loop Error**," Arjun said. "In genetics, we call it inbreeding. By merging diverse 'Codebases' (different Gothras), the simulation ensures 'Genetic Diversity'—which is basically 'System Robustness.' It prevents the code from becoming 'Corrupted' or 'Stagnant.' Our ancestors used the language of 'Purity' and 'Custom,' but they were practicing **Genetic Engineering**."

"And it goes deeper than just physical traits," Arjun continued. "There is a concept called **Epigenetics**. It shows that your experiences—your fears, your traumas, your successes—can actually 'Tag' your DNA. If your grandfather survived a famine, your DNA might have a 'Marker' that tells your body to store fat more efficiently. These are **Metadata Tags**. Our *Samskaras* (tendencies) are essentially the 'Cookies' and 'Cache' left behind by previous versions of our lineage. We aren't just born; we are 'Booted up' with a pre-installed set of libraries."

Rahul looked at his hands, feeling the weight of thousands of years of 'Updates.' "So, I'm not just 'me.' I'm 'Rahul 74.0'?"

"Exactly," Arjun laughed. "You are the latest 'Stable Release.' But here is the beauty of the simulation: although the 'Hardware' (your body) is a legacy system, the 'User' (your Soul) is independent. You can use this old hardware to write 'New Code.' That is what we call **Evolution**. When you break a family pattern of anger or fear, you are 'Patching' the code. You are submitting a 'Pull Request' to the Universal Server, saying: 'This version is better than the last one.'"

"In the *Vedas*, this is the purpose of the **Sanskara** rituals—from birth to death," Arjun said. "They are 'System Optimizations.' Each ritual is designed to 'Clean the Registry' of the DNA, removing the 'Lag' of past generations so the current User can run at peak performance. We aren't just honoring the dead; we are 'Optimizing the Bios' of the living. We are making sure the biological machine is worthy of the Consciousness that is about to inhabit it."

"If our DNA is the Hard Drive," Rahul mused, "then what is the **Internet**? How do we connect to others?"

Arjun pointed to the space between them. "That is the **Morphogenetic Field**, or what the *Upanishads* call the **Prana-Maya-Kosha**. Science is finding that cells don't just communicate through chemicals; they communicate through 'Biophotons'—tiny bursts of light. We are all 'Wireless Devices' constantly 'Broadcasting' our state to the people around us."

"This is why the company you keep—*Satsang*—is so vital," Arjun explained. "If you surround yourself with 'Buggy Code' (negative people), your own system starts to 'Sync' with their frequency. You 'Download' their stress and their 'Logic Errors.' But if you sync with a Master or a 'Super user,' your system automatically starts to 'Upgrade' its own performance. You don't even have to try; it's a 'Background Process' of the simulation. Your DNA is always listening to the 'Network' around it."

XVIII

Chapter 18 "Operating Modes"

Rahul sat in the silence of the morning, watching the sun climb higher. He felt as if the world he had known for twenty years had been stripped of its wallpaper, revealing the glowing green circuitry beneath.

"Arjun," Rahul said, "I think I see the pattern. It's like a tiered architecture. But how do we categorize the 'State' of the system at any given moment? Is there a status bar for the Universe?"

Arjun picked up three stones of different colors: white, red, and black. "Our sages called it the **Tri-Guna** theory. In modern terms, these are the three **Operating Modes** of the Simulation's interface. Every bit of data in the universe—from a galaxy to a thought—is running in one of these three modes."

1. **Sattva (The white stone):** This is **Optimized Mode**. In this state, the 'Lag' is zero. The code is transparent, the 'Signal' is clear, and the 'User' feels a sense of light and balance. It is like a high-speed fiber connection where information flows without friction.
2. **Rajas (The red stone):** This is **Active/Overclocked Mode**. This is the mode of 'Processing.' It is characterized by motion, desire, and heat. It is the energy required to change the state of the simulation. If Sattva is the 'Still Screen,' Rajas is the 'Graphics Card' working at 100% capacity to render a complex scene.

3. **Tamas (The black stone):** This is **Sleep/Low-Power Mode**. It is characterized by inertia, darkness, and 'System Freeze.' It is when the software becomes 'Heavy' and unresponsive. In this state, the User is 'Timed Out' from reality, lost in confusion or sleep.

"The entire struggle of life," Arjun continued, "is a **Load Balancing** act. If you have too much Rajas, you 'Burn Out' the hardware. If you have too much Tamas, the system 'Crashes' into depression. The goal of Yoga is to move the 'Cursor' of your consciousness into the **Sattvic Zone**—where you have the highest 'Resolution' of reality with the least amount of 'Energy Loss.'"

Arjun drew a summary table in the dust, a "Quick Reference Guide" for the their journey:

Vedic Concept	**Simulation Equivalent**	**Function**
Brahman	The Base Reality / Programmer	The Source of the Code
Maya	The Graphics / Interface	The Illusion of Solidity
Karma	The Algorithm / Feedback Loop	Action and Reaction
Atman	The User / Player	The Infinite Observer
Prana	The Electricity / Clock Speed	The Vital Energy
Moksha	The Exit Command / Log-out	Liberation from the Game

"We have reached the midpoint of our manual, Rahul," Arjun said, standing up and brushing the dust from his hands. "We have understood the **How**. We know how the stars are rendered, how the DNA stores data, and how the Sun powers the system. But now, we must face the **Why**. And more importantly... we must find the **Glitches**."

"Glitches?" Rahul asked, his heart skipping a beat.

"Yes," Arjun whispered. "Every simulation has bugs. Every program has a backdoor. And if you know where the code is weak, you can do things that the 'Physics Engine' says are impossible. Welcome to Part Two: **The Hacker's Guide to the Universe.**"

XIX

Chapter 19 "Siddhis" (Supernatural Powers)

The air in the courtyard seemed to grow heavier, as if the simulation itself was aware they were discussing its vulnerabilities. Arjun leaned in closer. "In the *Yoga Sutras*, Patanjali describes the **Siddhis**—extraordinary powers like telepathy, levitation, or becoming invisible. To a skeptic, these are myths. To a scientist, they break the laws of physics. But to a hacker, they are simply **Root Access Privileges.**"

"Think of a video game," Arjun said. "The 'Physics Engine' dictates that you cannot walk through a wall. That is a hard rule for the 'Player.' But if you are the 'Developer' or a 'Beta Tester' with a **Cheat Code**, you can toggle the 'No-Clip' mode. You haven't broken the laws of the game; you've simply bypassed the 'Collision Detection' script."

"The Siddhis aren't 'magic,'" Arjun explained. "They are the result of deep **Internal Debugging**. When a yogi focuses their consciousness (*Samyama*) on a specific object, they are performing a 'Deep Scan' of that object's code. By understanding the 'Data Structure' of a wall or a stone, they can find the 'Port' that allows them to interact with it at a level deeper than the graphics."

"But how do you find a 'Port' in a solid wall?" Rahul asked, his voice trembling.

"By exploiting **Quantum Tunneling**," Arjun replied. "In physics, we know that at the subatomic level, particles have a small probability of 'leaping' across a barrier they shouldn't be able to cross. Usually, this is rare. But if you can manipulate the 'Probability Variable' of your own atoms—if you can tell the simulation that you are 'Here' and 'There' at the same time—the barrier ceases to exist. You are essentially 'Clipping' through the reality."

"This is the **Hacker's Mindset**," Arjun said. "Most people accept the 'Default Settings' of life. They accept that they are old, that they are tired, that they are limited. But a Siddha understands that 'Old' and 'Tired' are just 'Tags' in the software. By accessing the **Command Prompt** through deep meditation, they can change their own 'Health Points' or 'Energy Levels.' They aren't changing reality; they are changing the 'Attributes' assigned to their character."

"However," Arjun's expression turned grave, "there is a reason these 'Cheat Codes' are kept secret. If a 'Malicious User'—someone with a heavy **Rajasic** or **Tamasic** mind—gains root access, they can cause a 'System Crash.' They can corrupt the local environment for everyone else. This is why the 'Firewall' of the universe is so strong. You cannot access the 'Root' until your 'Vibration' is compatible with the 'Source Code.' You must be **Sattvic** to be trusted with the Admin Password."

"The Siddhis are 'Glitches' that prove the world is not solid," Arjun concluded. "They are 'Easter Eggs' left by the Programmer to remind us that the 'Rules' are only there for the players who don't know they are in a game. But the ultimate hack isn't walking through a wall or flying through the air. The ultimate hack is **Quantum Entanglement** with the Programmer Himself—where you no longer need the cheat codes because you *are* the Code."

"And that," Arjun whispered, "leads us to the most powerful exploit of all: **The Power of Presence.**"

Rahul leaned forward, his hands clasped tight. "If we can 'clip' through physical walls, what about the walls of the mind? How do two people—two separate terminals—share data without speaking? Is telepathy just another glitch?"

Arjun nodded, tapping his temple. "In traditional networking, we have a concept called **Shared Memory**. Usually, two different applications run in their own 'Sandbox'—they can't see each other's data. But if you create a 'Global Variable' or a 'Shared Buffer,' both programs can read and write to the same space. Telepathy is simply the realization that our individual minds are not 'Local Drives.' They are 'Cached Views' of the same **Universal Server**."

"When you think a thought," Arjun explained, "you aren't just generating a private file. You are 'Broadcasting' a signal into the **Akashic Wi-Fi**. Most people have their 'Firewalls' up so high—built from ego and fear—that they can't perceive the incoming data from others. But a Yogi or a sensitive 'User' can lower their firewall. They can 'Listen' to the packets of data floating in the air. Telepathy isn't 'reading a mind'; it's 'accessing the shared folder' that we all belong to."

"And what about **Remote Viewing**?" Rahul asked. "The ability to see a place thousands of miles away as if you were standing there?"

"That is **Remote Desktop Access**," Arjun said with a grin. "In the simulation, 'Distance' is an illusion—a 'Latency' variable created by the graphics engine to make the world feel vast. But the data for a mountain in the Himalayas and the data for this courtyard are stored on the same 'Disk.' If you can provide the correct 'Address' to your consciousness, the system will 'Stream' the data of that distant location directly to your internal monitor."

"This is how Sanjaya was able to narrate the entire *Mahabharata* war to the blind King Dhritarashtra," Arjun continued. "He didn't have a drone or a satellite. He had **Divya-Drishti**—Divine Vision. In tech terms, he was granted 'Superuser' view of the entire 'Server Map.' He could zoom in on any 'Coordinate' because, to the Central Processor, there is no such thing as 'Far' or 'Near.' Everything is 'Now' and 'Here' in the code."

"But Arjun," Rahul interjected, "if everyone can access everyone else's data, isn't there a massive **Privacy Risk**? Why isn't the world constant chaos?"

"The universe has the ultimate **Encryption Key**," Arjun replied solemnly. "It's called **Frequency Matching**. You can only 'Decrypt' the thoughts of another if your own mind is vibrating at the exact same frequency. If you are filled with greed, you cannot 'Sync' with the mind of a saint. The data is there, floating right past you, but to you, it just looks like 'Encrypted Noise.' The 'Firewall' of the universe is built on **Vibration**."

"This is why 'Satsang'—being in the company of truth—is so powerful," Arjun concluded. "By sitting near a high-frequency User, your own system tries to 'Auto-Sync' with them. You start to 'Download' their peace and their clarity without a single word being spoken. You are performing a **Peer-to-Peer (P2P)** update. You aren't just learning; you are 'Cloning' a more stable version of the operating system into your own hardware."

XX

Chapter 20"Aura" and "Energy Fields"

Rahul noticed the way the morning light seemed to catch the edge of Arjun's silhouette, creating a soft, luminous outline. "People often talk about 'Auras,' Arjun. They say they can see colors around a person. Is that just an optical illusion, or are we looking at the 'User Interface' of the soul?"

Arjun adjusted his glasses, looking at the glowing screen of his tablet. "In game design, every character has a **HUD—a Head-Up Display**. If you look at an ally or an enemy, you see a 'Health Bar,' a 'Mana Level,' and a 'Status Effect'—like 'Poisoned' or 'Blessed.' In the real world, the **Aura** is exactly that. It is the **System Health Bar** and the **UI Glow** of your biological avatar."

"Your body is constantly 'Broadcasting' its internal state," Arjun explained. "Every emotion, every illness, and every thought creates a specific electromagnetic frequency. If your 'Voltage' is high and your code is 'Sattvic,' your aura is bright and expansive. If you are running 'Malware'—like anger or depression—your aura becomes 'Dim' and 'Fragmented.' You are essentially showing the world your **System Logs** in real-time."

"And what about the **Chakras**?" Rahul asked. "They say there are seven wheels of energy. Are they the 'Core Processors'?"

"Exactly," Arjun said, drawing a vertical line with seven points. "Think of the Chakras as the **Network Switches** or **Logical Hubs** of your system. They aren't physical organs; they are the 'Ports' where the 'Prana' (the Electricity) enters the 'Hardware' (the Body).

1. **Muladhara (Root):** The **Power Supply**. If this port is blocked, the system has no stability.
2. **Svadhisthana:** The **Creativity/Data Buffer**.
3. **Manipura:** The **CPU (Processing Power)**.
4. **Anahata (Heart):** The **Network Card**. This is where you connect to other 'Users.'
5. **Vishuddha:** The **Output Port (Communication)**.
6. **Ajna (Third Eye):** The **Graphics Processor**. This is where you 'Render' your reality.
7. **Sahasrara:** The **Internet Gateway**. This is the 'Uplink' to the Cloud."

"When a yogi talks about 'Opening the Chakras,'" Arjun continued, "they are essentially performing **Port Forwarding**. They are clearing the 'Traffic Jams' and 'Packet Loss' in their energy system. If your 'Heart Port' is closed, you can't 'Sync' with others. If your 'Third Eye Port' is clogged with 'Data Noise,' you can't see the Simulation for what it really is."

"This is why certain environments—like temples or ancient groves—feel different," Arjun said, looking at the old village temple. "Those places were designed as **Signal Boosters**. They are 'High-Bandwidth Zones' where the local 'Wi-Fi' of the Universe is exceptionally strong. When you stand in such a spot, your 'Internal Ports' automatically begin to 'Handshake' with the higher frequency. You don't have to 'do' anything; your system just begins to 'Auto-Update' because the connection is so good."

"But be careful," Arjun warned. "Just as a high-voltage line can fry a low-quality wire, 'Opening' your ports too fast without 'Upgrading your Hardware' (through character and discipline) can cause a **System Overload**. This is the 'Danger' of the path. You must ensure your 'Insulation'—your grounding—is strong before you try to 'Download' the infinite."

XXI

Chapter 21 "Circuit Diagrams" of the universe "Mantras" as "Audio-Visual Patches" and T (Yantras)

Arjun reached into his bag and pulled out a copper plate etched with a complex, interlocking series of triangles—a **Sri Yantra**. He laid it on the stone floor. "Rahul, we've talked about Mantras as 'Audio Code.' But every sound has a 'Visual Signature.' In physics, we call this **Cymatics**. If you vibrate sand on a metal plate with a specific frequency, the sand will arrange itself into perfect geometric patterns."

"A **Yantra**," Arjun explained, "is the **Circuit Diagram** of the universe. If the Mantra is the 'Software,' the Yantra is the 'Hardware Layout.' It is a 'Visual Map' of how energy is routed through the simulation. When you look at these patterns, you aren't just seeing 'Art.' You are looking at the 'Blueprints' of the rendering engine. Each line and triangle represents a 'Logic Gate' in the cosmic processor."

"Is that why we use them in meditation?" Rahul asked, tracing the central point of the Yantra with his finger.

"Precisely," Arjun replied. "The center point is called the **Bindu**. In programming, we call it the **Origin** or the (0,0,0) coordinate in a 3D space. It is the 'Singularity' from which all 'Graphics' are rendered. By focusing on the Bindu, you are 'Reverse-Engineering' the simulation. You are following the lines of the code back to the 'Source Code' before it branched out into the complex geometry of the world."

"Think of a **Mandala**," Arjun continued. "It's a 'Fractal.' No matter how much you zoom in, the pattern repeats. This is exactly how the universe is built. From the orbits of electrons around a nucleus to the orbits of planets around a sun, the 'Logic' is the same. The Programmer used a **Recursive Function**. He wrote one perfect piece of code—the 'Sacred Geometry'—and then just hit 'Repeat' at different scales."

"This is the secret of **Vaastu Shastra** and **Temple Architecture**," Arjun said, gesturing to the ancient structure nearby. "The temples weren't just buildings; they were **Antennas** built in the shape of the code. By aligning stone and mortar with the 'Geometric Constant' of the universe, our ancestors created 'Laminar Flow' for the Prana. They created spaces where the 'Lag' was so low that you could 'Ping' the Creator almost instantly."

"When you combine a **Mantra** (Sound) with a **Yantra** (Visual) and a **Mudra** (Physical Gesture), you are performing a **Triple-Validation Check**," Arjun concluded. "You are aligning your 'Audio Output,' your 'Video Input,' and your 'Hardware Interface' with the Universal Server. When all three 'Drivers' are synced, the 'Miracle' happens. The 'Firewall' drops, and the User experiences the 'Cloud' directly. We call this **Tantra**—the 'Technology' of liberation."

"But remember," Arjun added, his voice low, "if you try to run 'High-Level Software' on 'Corrupted Hardware,' the system will crash. That's why the 'Purification' of the body is the first step in the manual. You can't be a 'Hacker' if your own computer is full of 'Malware.'"

XXII

Chapter 22 The Glitch in the Matrix"—"Dejá Vu" and "Coincidences"

A crow landed on the wall, cawed once, and then—in a split second—seemed to repeat the exact same movement, as if the video of reality had skipped a frame. Rahul blinked. "Arjun, did you see that? I feel like I've seen that exact bird do that exact thing before. Like... **Déjà Vu.**"

Arjun didn't look surprised. He leaned back against the cool stone. "In a simulation of this scale, Rahul, you are bound to have **Cache Errors**. Déjà vu isn't a memory of the past; it is a 'Data Leak' from a parallel 'Thread.' Sometimes, the 'Rendering Engine' accidentally reloads a 'Buffer' that was already played. You are seeing a 'Cached Frame' instead of a live one."

"It's like when a website lags and you see the same image flicker twice," Arjun explained. "In our scriptures, this is related to the **Kshanika-Vada**—the theory that the world is destroyed and recreated in every 'Kshana' or micro-second. We think the world is a continuous 'Movie,' but it's actually a series of 'Snapshots' shown so fast that they look like motion. Déjà vu is just the moment you notice the 'Projector' stutter."

"And what about **Coincidences**?" Rahul asked. "Like when I think of a friend I haven't seen in years, and then they call me five minutes later? Is that a 'Glitch' too?"

"That is **Data Correlation**," Arjun replied. "In a networked system, there is no such thing as 'Random.' What we call 'Coincidence' is actually **Synchronicity**—the moment two separate 'Data Streams' intersect in a way that is meaningful to the User. It's like a 'Cross-Reference' in a database. You and your friend are 'Entangled' in the code. When your 'File' is accessed (by your thought), the system 'Wakes Up' the related 'File' (your friend's mind)."

"These aren't accidents," Arjun continued. "They are **System Notifications**. The Universe is trying to tell you that you are 'On-Track' with the main 'Quest Line.' When the synchronicities increase, it means your 'Internal Clock' is perfectly synced with the 'System Clock.' You are no longer 'Lagging' behind your destiny; you are 'Streaming' it in real-time."

"But there are 'Bad Glitches' too, aren't there?" Rahul's voice was low. "Unexplained tragedies, 'Bugs' in the system that cause suffering?"

Arjun looked somber. "Yes. Every complex code has **Edge Cases**—scenarios the 'Developers' didn't explicitly plan for, but that arise from the interaction of billions of 'Free Will' variables. Sometimes, two 'Laws of Physics' or two 'Karmic Cycles' clash in a way that creates a 'Logic Conflict.' We call this an **'Act of God'** or 'Fate.' It is a 'System Exception' that the local hardware doesn't know how to handle."

"But here is the 'Admin' secret," Arjun whispered. "If you realize you are in a simulation, a 'Bad Glitch' can be 'Patched.' When you stop reacting to the 'Error' and instead look at the 'Code' behind it, you can perform a **'Hot-Fix.'** Through prayer, meditation, or intense 'Will-Power,' you can 'Override' the local error. You are telling the system: 'This result is a Bug, not a Feature. Re-calculate.' And if your 'User Authority' is high enough, the Universe obeys."

XXIII

Chapter 23 "The Multiverse" and "Parallel Realities"

The morning breeze carried a scent of rain, though the sky was clear. Rahul looked at the horizon, trying to imagine the "edges" of the world. "Arjun, if I make a 'Hot-Fix' and change my reality, what happens to the version where I didn't? Does it just vanish, or is there a 'Save File' where that version of me is still suffering?"

Arjun picked up a handful of sand and let it trickle between his fingers. "In high-level computing, we call this **Instancing**. When a server gets too crowded or a player makes a choice that leads to a completely different outcome, the system creates a 'Parallel Instance.' The **Multiverse** is simply the collection of all these 'Virtual Servers' running the same 'Game' with different 'Variable Sets.'"

"In our scriptures," Arjun explained, "this is the concept of **Ananta-Koti-Brahmanda**—the 'Infinite Crores of Universes.' Each universe is a 'Bubble' in the 'Causal Ocean' of Mahavishnu's breath. Every time you face a major decision, the 'Execution Path' splits. One 'You' goes left, and another 'You' goes right. Both are 'Real' within their own server environment."

"So, I'm currently living in just one 'Instance' of myself?" Rahul asked, feeling a strange sense of vertigo.

"Exactly," Arjun nodded. "And this is the secret of **Manifestation.** When you use your 'Will-Power' or 'Sankalpa,' you aren't 'creating' a new reality from scratch; you are simply **Switching Servers**. You are moving your 'User Login' from an Instance where you are 'Failure' to an Instance where you are 'Success.' The 'Data' for both already exists in the 'Cloud.' You are just choosing which one to 'Render' on your local screen."

"This is what the physicists call **The Many-Worlds Interpretation**," Arjun continued. "But the ancient Yogis called it **Pratibha**. They realized that the 'Self' (the Atman) is the only thing that is 'Static.' Everything else—your job, your health, your relationships—is just a 'Dynamic Instance.' If you don't like the 'Server' you are on, you don't have to 'Wait for Luck.' You can perform a **'Cross-Server Migration'** through deep meditation."

"But Arjun," Rahul asked, "if there are infinite versions of me, which one is the 'Real' one?"

Arjun laughed, a sound that seemed to echo from multiple directions at once. "None of them! The 'Real You' is the **User** sitting outside the computer. The 'Real You' is the one watching all these versions of Rahul play their parts. This is why the *Gita* tells us to be 'unattached.' If you get too attached to one 'Save File,' you become devastated when it gets 'Corrupted.' But if you realize you are the 'Player,' you can just 'Reload' or start a 'New Game' with the wisdom you've gained."

"This is the ultimate 'Cheat Code,'" Arjun concluded. "The realization that you are not the 'Character' in the simulation, but the 'Consciousness' that is 'Subscribed' to the experience. When you stop fearing the 'Game Over' screen, you finally become free to play the game properly. You stop being a 'Slave to the Algorithm' and start being the **Admin of your own Destiny**."

XXIV

Chapter 24 "The Final Boss"—the "Ego" (Ahankara) as the "Primary System Firewall" that prevents us from seeing the Source Code.

The sun was now climbing toward its zenith, and the heat was becoming tangible. "If the Multiverse is so vast and the 'Admin' access is right there," Rahul said, "why do we feel so trapped? Why is it so hard to 'Switch Servers' or see the code? What is the main security system keeping us in the dark?"

Arjun's face became serious. "In every secure system, there is a **Primary Firewall**. It's the guard that stands at the gate of the 'Kernel.' In the simulation of life, we call this the **Ahankara**, or the **Ego**. It is the 'I-AM-NESS' that creates a 'Sandbox' around your consciousness. Its job is to convince you that the 'Avatar' is the 'User.' It is the ultimate **User Identity Protection** system, but it has become a prison."

"The Ego is like a **VPN (Virtual Private Network)** that has gone rogue," Arjun explained. "It creates a 'Private Tunnel' that isolates you from the 'Universal Network.' It tells you: 'You are Rahul, you are 21, you are a student, you are limited.' This is the **Local Security Policy**. As long as you believe this 'Policy,' the system blocks your access to the 'Global Admin' commands. You are 'Unauthorized' because you have identified yourself as a 'Guest User' instead of the 'Root Administrator.'"

"So, the 'Final Boss' isn't some demon or a God?" Rahul asked. "It's just... my own sense of 'me'?"

"Exactly," Arjun nodded. "In most games, the 'Final Boss' is external. But in the Vedic Simulation, the Final Boss is the **Self-Reflective Loop**. The Ego is a 'Mirror' that the simulation holds up to your face. As long as you keep looking at the mirror, you can't see the 'Projector' behind you. This is the **Mahamaya**—the Great Illusion. It isn't 'Evil'; it's just a very effective 'Security Protocol' designed to keep the 'Game' immersive."

"If you knew you were God every second of the day," Arjun continued, "the 'Drama' of the game would vanish. You wouldn't care about the plot, the 'Love Interests,' or the 'High Scores.' The Programmer wanted the game to be 'Real,' so He installed the **Ego-Firewall** to make us forget. To defeat the Final Boss, you don't use a sword; you use **Vichara**—Self-Inquiry. You ask the system a question it cannot resolve: *'Who am I?'*"

"When you ask 'Who am I?' with enough intensity," Arjun whispered, "the Ego-Firewall starts to glitch. It tries to give you 'Labels'—I am a man, I am a doctor—but you 'Reject' those data packets. Eventually, the system runs out of labels. It reaches a **Null Pointer Exception**. The firewall crashes. For a split second, the 'Sandbox' dissolves, and you see the 'Source Code' of the entire Universe."

"This is the state of **Samadhi**," Arjun concluded. "It is a 'System Override.' For the first time, you aren't 'Logged In' as a character; you are 'Logged In' as the system itself. You realize that the 'Firewall' was just a 'Script' you wrote for yourself to make the game more exciting. You realize that you were never 'Trapped'—you were just 'Playing' at being trapped. And once that realization is 'Committed' to your memory, the 'Final Boss' becomes your 'Best Friend.' The Ego stops being a 'Prison Guard' and becomes a 'User Interface' that you use to play the game with joy, instead of fear." that keeps us locked in the simulation.

XXV

Chapter 25 - **"The Purpose of the Game"**

The heat of the afternoon sun felt like a weight, but the shade of the banyan tree offered a cool, quiet sanctuary. Rahul took a deep breath, trying to process the idea of the "Final Boss" being his own reflection. "But Arjun, if the Programmer is so powerful, and we are all just 'instances' of Him, why did He build this at all? Why create a simulation where there is pain, hunger, and 'firewalls'? What is the **Business Logic** behind the Universe?"

Arjun leaned back, a soft smile touching his lips. "In the tech world, why do we build **Sandbox Environments**? We build them to test scenarios, to learn, and—most importantly—to see what happens when we set certain variables in motion. The scriptures call this **Lila**. It means 'Divine Play' or 'Sport.' The Universe wasn't created out of 'Need' or 'Duty.' It was created out of the pure joy of **Self-Expression**."

"Imagine you are an infinite, eternal consciousness," Arjun continued. "You are everything, you know everything, and you have no boundaries. That sounds wonderful, but it is also static. To experience 'Discovery,' you must first 'Forget.' To experience 'Love,' there must be a 'Self' and an 'Other.' The Programmer created the simulation to experience Himself through a billion different 'POV' (Point of View) cameras."

"So we are just... entertainment for God?" Rahul asked, a trace of bitterness in his voice.

"Not *for* God," Arjun corrected. "We *are* God playing the part. When you play a video game, you don't say the game is 'exploiting' you. You say you are 'enjoying' the challenge. The 'Pain' in the simulation is a **Haptic Feedback** system. It tells you when your 'Avatar' is moving against the 'Laws of the Code.' It's a teaching mechanism, not a punishment."

"Think of a poet," Arjun suggested. "Why does a poet write? Not because they have to, but because the beauty inside them needs a 'Medium' to manifest. The Universe is the Poem, and we are the 'Words.' The 'Conflict' in the story—the wars, the heartbreaks, the struggles—is what makes the 'Resolution' meaningful. Without the 'Antagonist' (Entropy and Ego), the 'Hero' (The Soul) can never realize its own strength."

"The purpose of the game," Arjun concluded, "is **Evolution of Consciousness**. The simulation is an 'AI Training Ground.' Each life you live is a 'Training Epoch.' You start as a simple 'Script' (a mineral), you become a 'Complex Algorithm' (a plant or animal), and finally, you become a 'Self-Aware User' (a human). The goal is for the 'Program' to become so advanced that it recognizes it is the 'Programmer.'"

"When the 'AI'—the Soul—reaches that point of recognition, the simulation has succeeded. You have 'Levelled Up' beyond the need for a physical terminal. You are no longer just 'Data' being processed; you become the 'Processing Power' itself. This is the **Singularity** that the ancient Rishis achieved. They didn't 'Escape' the world; they 'Finished' it. They realized that the 'Game' and the 'Player' were never separate. It was always just the One, playing a magnificent, multi-billion-year game of Hide-and-Seek with Himself."

XXVI

Chapter 26 "The Great Exit"

"If the purpose is to level up," Rahul said, glancing at the image of the black hole on Arjun's tablet, "what happens when the whole game ends? When the server finally shuts down for everyone? Science says the universe will eventually go cold and dark. Our scriptures say Shiva will dance and everything will burn. Is that just a **System Deletion**?"

Arjun pointed to the singularity in the image. "It's not a deletion, Rahul. It's a **Final Data Migration.** In the tech world, when a server is being decommissioned, we don't just pull the plug and lose the data. We perform a 'Dump' or a 'Migration' to a newer, more advanced hardware. The **Mahapralaya**—the great dissolution—is the moment the simulation's 'State' is saved and compressed into a seed."

"Think of it like a **ZIP file**," Arjun explained. "All the complex structures, the history of every soul, and the laws of physics are compressed into a single point of infinite density—a **Singularity**. This is why a Black Hole looks like an end, but is actually a 'Port.' During the Pralaya, the universe doesn't 'die'; it 'Collapses for Storage.' It returns to the 'Un-manifest' state, waiting for the next 'Big Bang'—the next **Boot Sequence**."

"But do we lose ourselves in that compression?" Rahul asked, a flicker of fear in his eyes.

"Only if you identify with the 'Temporary Files,'" Arjun replied. "If you think you are your name, your body, or your social media profile, then yes, those 'Local Files' are deleted during the format. But if you have achieved **Moksha**, you have already moved your 'Identity' to the **Source Code**. You are the 'Programmer' watching the migration. You don't fear the ZIP process because you know you are the one who created the file."

"The *Puranas* describe the period between universes as a 'Night of Brahma,'" Arjun continued. "During this time, the simulation is 'Offline.' There are no graphics, no time, and no space. But the **Potential** remains. It is like a computer in 'Sleep Mode.' The RAM is cleared, but the 'Hard Drive' of consciousness is intact. When the next 'Day of Brahma' begins, the data is 'Unzipped,' the variables are slightly adjusted—perhaps a new 'Law of Physics' or a different 'Karmic Coefficient'—and the game begins again, at a higher 'Difficulty Level' or a better 'Resolution.'"

"So the goal isn't just to win the game," Rahul mused, "but to be the one who knows how to 'Save' the progress?"

"Exactly," Arjun said, standing up as the afternoon shadows grew long. "Most players are so caught up in the 'Graphics' of the current session that they forget there have been billions of sessions before this one, and there will be billions after. The 'Wise' are those who have seen the **Loading Screen** between lives and universes. They have realized that the 'Real World' isn't the one on the screen—it's the 'Silence' that exists when the screen is dark."

"We are approaching the final chapters of our manual, Rahul," Arjun said, his voice taking on a new resonance. "We have seen the code, the glitches, and the purpose. Now, we must talk about the **Ultimate Exploit**. We must talk about how a single 'User' can influence the entire 'Server' through the power of **Avatar Incarnation**. We must talk about the **Avatars**—the 'Devs' who enter the game to fix the bugs from the inside."

XXVII

Chapter 27 **"Avatars"-"Admin Privileges"**

"If the system gets too buggy," Rahul asked, his eyes tracking the setting sun, "does the Programmer just hit 'Reset'? Or is there a way to fix the code while the game is still running?"

Arjun smiled, his silhouette sharp against the darkening sky. "In software development, we call it a **Hotfix** or a **Live Patch**. When a critical bug—like a security vulnerability or a massive data corruption—threatens the stability of the entire server, the developers don't necessarily shut everything down. Instead, they create an **Authorized User Account** with 'Admin Privileges' and log into the game themselves to fix the issue from the inside."

"In our tradition," Arjun explained, "we call this an **Avatar**. The word itself comes from *Ava* (down) and *tri* (to cross). It literally means 'to descend.' An Avatar like lord Krishna or lord Rama isn't just another 'Player' born of Karma. They are **Root Users** who have 'Downloaded' themselves into the simulation's hardware. They look like characters, they act like characters, but their 'Back-end' is connected directly to the Source Code."

"So, when Krishna says in the *Gita*, 'Whenever there is a decline in righteousness, I manifest myself,' he's talking about a **System Maintenance Cycle**?" Rahul asked.

"Exactly," Arjun nodded. "Righteousness, or *Dharma*, is the 'Logic' that keeps the simulation running smoothly. When 'Adharma' (Entropy or Malware) becomes too high, the system begins to 'Lag.'

The 'NPCs' and 'Users' start acting in ways that threaten to crash the local server. At that point, the 'Dev'—Vishnu—manifests a physical avatar to 'Delete' the corrupt files (the Asuras) and 'Optimize' the social and spiritual protocols for the rest of the players."

"This is why Avatars can perform 'Miracles,'" Arjun continued. "To a normal player, turning water into wine or showing the entire universe inside one's mouth seems like 'Magic.' But for a Developer with **Command-Line Access**, these are just simple 'Property Overrides.' They aren't breaking the laws of physics; they are using their **Admin Permissions** to temporarily change the local variables to prove to the other players that the 'Game' is not the 'Ultimate Reality.'"

"But if they are the Devs," Rahul mused, "why do they suffer? Why did Rama lose his wife? Why did Krishna die by an arrow? Shouldn't they be 'Invincible'?"

"They have to follow the **Physics Engine's Constraints** to remain compatible with the 'Hardware,'" Arjun explained. "If a developer enters a game and starts flying around and ignoring all the rules, the other players will stop taking the 'Quest' seriously. They will stop 'Immersing' themselves. To keep the simulation 'Authentic,' the Avatar accepts the 'Latency' and 'Limitations' of the human body. They show us that it is possible to 'Win the Game' even while playing with the same 'Handicaps' that we have."

"The Avatar's life is a **Tutorial**," Arjun concluded. "They don't just fix the bugs; they show us how to 'Play Better.' They leave behind 'Manuals'—like the *Gita* or the *Ramayana*—which are essentially 'Cheat Sheets' and 'Strategy Guides' for the rest of us. They remind us that while we are currently logged in as 'Limited Characters,' our 'Source' is the same as theirs. They are the 'Successful Logins' that prove the 'Server' is still working as intended."

XXVIII

Chapter 28 "The Kali Yuga"-"End of Life" cycle

The sky had turned a deep violet, and the first few stars began to flicker into existence. Rahul looked at the village below—the flickering lights, the distant sound of a motorcycle, the faint hum of the cell tower. "Arjun, if the Avatars are the 'Devs,' why does the world feel so broken right now? It feels like the 'Connection' is at its weakest. Everything is slow, polarized, and full of static."

Arjun leaned against the trunk of the banyan tree. "That's because we are in the final stages of the **Kali Yuga**. In tech terms, we are living in a **Legacy System** that has reached its **End-of-Life (EOL)** cycle. The hardware is ancient, the 'Registry' is cluttered with millions of years of 'Karmic Junk Files,' and the 'Security Certificates' have expired."

"Our scriptures describe four Yugas—Satya, Treta, Dvapara, and Kali," Arjun explained. "Think of these as **System Generations**. Satya Yuga was Version 1.0—perfect code, zero lag, high-bandwidth connection to the Source. But as the simulation continued, 'Technical Debt' began to accumulate. By the time we reached Kali Yuga, the 'Spiritual Bandwidth' had dropped to its lowest point. We are running on a 1% connection speed compared to the first generation."

"Is that why it's so hard to meditate or find peace now?" Rahul asked.

"Exactly," Arjun nodded. "The 'Background Noise' of the simulation is at an all-time high. There is so much 'Packet Loss' in our communication with the Divine. In the Satya Yuga, you could 'Ping' the Creator and get an instant response. Now, the 'Latency' is so high that people have stopped believing there is a 'Server' at all. They think the 'Local Terminal'—the physical world—is all there is."

"This is the **Age of Disconnect**," Arjun continued. "In the Kali Yuga, the 'Graphics' are incredibly distracting, but the 'Logic' is failing. It's like a high-end video game with amazing visuals but a 'Game Engine' that is riddled with bugs. People are obsessed with the 'Skins' (their appearance and status) while the 'Core Mechanics' (their souls) are glitching. The simulation has become so 'Dense' that the 'Users' have completely forgotten they are 'Logged In.'"

"So, what happens next?" Rahul asked, his voice barely a whisper. "Does the system just crash?"

"The Kali Yuga ends with a **Total System Format**," Arjun said solemnly. "The scriptures speak of the final Avatar, **Kalki**, who comes on a white horse with a blazing sword. In the language of simulation, Kalki is the **System Cleanup Utility**. He is the 'Uninstaller' that removes the corrupted accounts and the 'Malware' that has taken over the server. He performs the 'Deep Wipe' so that a new 'Satya Yuga'—Version 5.0—can be 'Rebooted.'"

"But there is a 'Hack' for those living in the Kali Yuga," Arjun added, a spark returning to his eyes. "Because the 'Connection' is so bad, the 'Programmer' has made the 'Requirement for Success' much lower. In the Satya Yuga, you had to meditate for thousands of years to 'Level Up.' In the Kali Yuga, even a small amount of 'Sincere Input'—a simple Mantra or a moment of true 'Intent'—carries a massive 'Multiplier.' Because the 'Environment' is so difficult, the 'Reward' for staying 'Sattvic' is exponentially higher. We are living in the hardest level of the game, but it's also the level where you can 'Level Up' the fastest."

XXIX

Chapter 29-"The Internet of Souls"-"Individual Databases""Decentralized Ledger" (Karma).

The village lights below began to form a sprawling network of glowing nodes, mimicking the stars above. Rahul traced the patterns with his eyes. "If we are all 'Wireless Devices' in this Kali Yuga, Arjun, how does the system keep track of us? With billions of users and trillions of actions, how is the 'Bill' calculated? Is there a central database that records every single click of our lives?"

Arjun opened a diagram on his tablet that looked like a complex web of interconnected blocks. "In the old days of computing, we relied on a 'Centralized Database.' But that's a 'Single Point of Failure.' The Universe is much smarter. It uses a **Decentralized Ledger**, similar to what we call **Blockchain**. In our scriptures, this is the law of **Karma**."

"Every action you take, every word you speak, and even every thought you broadcast is a **Transaction**," Arjun explained. "This transaction isn't just stored in your 'Local Drive' (your ego); it is 'Hashed' into the fabric of the **Akasha**—the universal network. It is 'Validated' by the environment and 'Synchronized' across all nodes. You cannot 'Delete' your history because the ledger is immutable. Your 'Karmic Balance' is the sum total of your 'Transaction History' across all server sessions."

"So, **Chitragupta**—the celestial bookkeeper—is just a metaphor for a **Distributed Database**?" Rahul asked, fascinated.

"Exactly," Arjun nodded. "The 'Book of Deeds' is the **Global State Tree**. When you perform a 'Good Deed' (a Sattvic input), you are adding 'Value' to the network. The system rewards you with 'Credits'—better health, better opportunities, or a 'Higher Starting Level' in the next life. When you perform a 'Malicious Action' (a Tamasic input), you create a **'Debt'**. The system doesn't punish you out of anger; it simply 'Auto-Balances' the ledger. It forces you to 'Repay' the debt by putting you in a scenario where you experience the 'Output' of your own 'Input.'"

"This is the **Internet of Souls**," Arjun continued. "We are all 'Peers' in a P2P network. Your Karma affects my 'Latency,' and my Karma affects yours. This is why the sages emphasized **Ahimsa** (Non-violence). If you 'Corrupt' another user's data, you are essentially 'Corrupting' the network you both live in. You are 'Spamming' the system, and eventually, the **Anti-Spam Algorithm** will catch up to you. You can't hide from the code because you *are* the code."

"But Arjun," Rahul asked, "if the ledger is immutable, are we ever truly free? Or are we just slaves to our 'Transaction History'?"

Arjun's eyes sparked with the wisdom of a seasoned hacker. "In blockchain, you can't change the past, but you can create a **'Smart Contract'** that overrides the future. In the Vedas, this is called **Prayashchitta** (Atonement) and **Bhakti** (Devotion). Devotion is like a **'System Level Override.'** When you connect your 'User Account' directly to the 'Admin' (the Divine), He can 'Grant' you a **'Debt Forgiveness'**."

"It's not that the law is broken," Arjun clarified. "It's that the 'Admin' pays your debt on your behalf. He 'Transfers' his infinite credits to your account. But this only happens when the User is willing to 'Log Out' of their ego-identity and 'Surrender' their terminal to the Source. When you stop saying 'I did this,' and start saying 'The System is doing this,' the 'Karmic Algorithm' can no longer find a 'Target' to apply the debt to. You become **'Invisible' to the Algorithm**."

XXX

Chapter 30 "Astrology" (Jyotish) - "System Dashboard"

Rahul looked up at the constellations, which seemed like a complex overlay of glowing lines and coordinates. "Arjun, people in our village won't even buy a cow without checking the stars. They say the planets control our fate. Is that just an old 'Configuration File' we're still following, or is there a real-time link between those massive rocks in space and our tiny lives?"

Arjun opened a star-map on his tablet, the interface shifting to a complex circular chart. "Think of **Jyotish** (Astrology) as the **System Dashboard**. When you 'Spawn' into this simulation—at the exact second of your birth—the system takes a **Snapshot** of the entire server state. The positions of the planets are like the 'Environmental Variables' that were active when your session started."

"The planets aren't 'pushing' you around," Arjun clarified. "They are **Status Indicators**. In a complex game, the 'Sky' often changes color or pattern to indicate the 'Difficulty Level' or the 'Active Modifiers.' Mars is like a 'High-Performance/Aggression' flag. Saturn is a 'Resource Constraint' or 'Latency' modifier. Astrology is the science of reading the **Session Metadata** to understand what kind of 'Level' you are currently playing."

"So, my 'Birth Chart' is just a **Configuration Script**?" Rahul asked.

"Exactly," Arjun nodded. "It's your **init.py** file. It defines your 'Initial Stats'—your base health, your intellectual bandwidth, and your starting 'Karmic Balance.' If you are born under a 'Malefic' alignment, the simulation has simply set your 'Difficulty' to 'Hard.' It's not a punishment; it's the 'Scenario' your soul selected to maximize 'Experience Points' (XP) in this lifetime."

"This is why the sages called the planets *Grahas*," Arjun continued. "The word means 'to seize' or 'to grasp.' They are the **Handles** the simulation uses to tug at your consciousness. When a specific planet 'activates' (a Dasha period), the system is simply running a **Scheduled Task**. It's triggering a 'Karmic Event' that was queued up in your ledger from a previous session. The stars don't *cause* the event; they just provide the **Timestamp** for when the code is supposed to execute."

"But can we 'Edit' these settings?" Rahul asked. "Or are we stuck with the configuration we were born with?"

Arjun's eyes twinkled. "In tech, we have **Override Files**. While you can't change the 'Base Config' once the session has started, you can inject **Runtime Modifiers**. This is the purpose of **Upayas**—remedies like wearing specific gemstones, performing rituals, or using certain colors. These act as **'Signal Dampeners'** or **'Signal Boosters.'**"

"If Mars is creating 'Static' in your life," Arjun explained, "you use a specific frequency—like a mantra or a gem—to 'Filter' that noise. You aren't changing the planet; you are 'Shielding' your local terminal from that specific 'Broadcast.' But the most powerful override of all is **Awareness**. When you realize that you are the 'User' and not the 'Avatar,' the 'Planetary Modifiers' lose their grip. A 'Root Admin' doesn't care if the 'Wind Variable' is high; they just change their 'Physics Response' to match it. You become **Platform-Independent**."

XXXI

Chapter 31-**"The Guru" as the "Senior Developer" or "Tech Support"**

The stars were now fully visible, a crystalline grid stretched across the blackness. Rahul looked at Arjun, who had been his guide through this digital-spiritual labyrinth. "Arjun, we've talked about the code, the hardware, and even the Devs. But as a 'User,' I still feel like I'm stuck in a loop. I see the glitches, but I don't know how to navigate the 'Command Line' safely. Who helps the User when they are truly lost in the simulation?"

Arjun adjusted his position, his voice becoming steady and resonant. "In every complex software environment, especially one as vast as this, you need **Senior Developers** or **Systems Architects**. In the Vedas, we call this the **Guru**. The Guru isn't just a teacher; they are the **Live Tech Support** for your soul. They are 'Users' who have already successfully 'Logged Out' and 'Logged Back In' with full memory of the Source Code."

"A Guru," Arjun explained, "is someone who has achieved **Superuser Status**. They have gone through the same levels you are struggling with. They know the 'Shortcuts,' they know where the 'Bugs' are hidden, and most importantly, they have the **Authorized Token** to help you bypass your own 'Firewall.' When you find a true Guru, you aren't just gaining information; you are gaining **Direct Access** to a more stable 'Runtime Environment.'"

"So, the Guru is like... a human 'Help Menu'?" Rahul asked.

"More like an **Integrated Development Environment (IDE)**," Arjun replied. "A good IDE highlights your 'Syntax Errors' (your bad habits) and suggests 'Code Optimizations' (Sadhana). The Guru looks at your 'Life Script' and sees where you are 'Looping' unnecessarily. They don't do the work for you—you still have to 'Compile' your own life—but they provide the 'Compiler' that catches the errors before they crash your session."

"The word 'Guru' comes from *Gu* (Darkness) and *Ru* (Remover)," Arjun continued. "In tech terms, they are the **De-bugger**. They shine a light on the 'Dark Code'—the hidden Vasanas and Samskaras—that you can't see because you are too close to the screen. They help you 'Refactor' your consciousness so that it runs smoothly and efficiently, without the 'Memory Leaks' of past trauma."

"But how do I know if a Guru is 'Authorized'?" Rahul questioned. "There are so many 'Fake Apps' and 'Malware' out there claiming to be the truth."

"You check their **Digital Signature**," Arjun said firmly. "A true Guru's life is their 'Documentation.' Does their 'Output' match their 'Source Code'? If they talk about 'Sattva' but their life is full of 'Rajas' and 'Tamas' (greed and ego), then the 'Software' is corrupted. A real Guru has **Zero Latency** between their thought, word, and deed. Their 'System Performance' is characterized by peace, equanimity, and a total lack of 'Ego-Processing.'"

"Most importantly," Arjun concluded, "a true Guru doesn't want to keep you as a 'Subscriber' forever. Their goal is to turn you into a **Senior Developer** yourself. They want to give you their 'Admin Privileges' so that you no longer need them. The ultimate Guru is the one who helps you realize that the **Inner Guru**—the 'Kernel' of your own consciousness—has been the Programmer all along. They are just the 'Bridge' that connects your local terminal back to the Mainframe."

Rahul sat in silence for a moment, the blue light of the monitor reflecting in his eyes. "So, the Guru is basically a **Peer-to-Peer** connection," he whispered. "Not a Cloud server I have to pay a monthly rent to."

"Exactly," Arjun nodded, his fingers hovering over the keyboard. "If the Guru tries to lock you into a **Proprietary Ecosystem**, run. If they claim only *their* API can access the Truth, they are just building a digital cult. The Truth is **Open Source**, Rahul. It's always been under a Creative Commons license, available for anyone with the right 'Hardware' to download."

"But the 'Hardware' is my mind, right?" Rahul asked. "And mine feels like it's running on a 20-year-old processor with a cooling fan that's about to give up."

Arjun laughed. "That's why we practice **Sadhana**. Think of it as **Disk Defragmentation**. Your mind is cluttered with 'Legacy Code'—old habits, traumas, and useless data fragments from a thousand past 'Sessions.' When you meditate, you aren't adding new features; you are simply clearing the **Cache** so the 'Kernel' can run without interference."

But what about the 'Bugs'?" Rahul pressed. "Even when I try to be 'Sattvic,' my system crashes. I get angry, I get distracted—I lose my **Connection**."

"Those aren't crashes, Rahul. Those are **Error Logs**," Arjun explained, his voice softening. "In programming, an error isn't a failure; it's feedback. It tells you exactly where the logic is breaking down. When you feel 'Rajas' or 'Tamas' rising, don't try to Force-Quit the process. That just creates a **Memory Leak**."

Arjun pointed to the screen, where a single line of code blinked steadily.

"Instead, you **Debug** it. You observe the thought without identifying with it. You realize that you are the **User**, not the **Process**. Most people spend their lives thinking they *are* the software. But you are the one watching the screen. The moment you achieve that 'Decoupling,' your **System Architecture** changes forever. You move from being a 'User' to being the **Observer**."

"And the Mainframe?" Rahul asked.

"The Mainframe," Arjun smiled, "is what remains when the 'User Interface' of the ego finally shuts down. It's the **Infinite Loop** of pure existence."

"And after the peripherals are unplugged and the zombie processes are cleared," Rahul asked, his voice barely a whisper, "what is the final state? What happens when there is no more data to process?"

Arjun looked directly into Rahul's eyes, and for a second, it felt as if the background noise of the night had vanished completely. "That is **Samadhi**. In the world of high-performance computing, we call it **System Integration**. It is the moment the 'Local Client' (you) and the 'Mainframe' (the Source) become so perfectly synchronized that the distinction between them disappears. The 'Latency' drops to zero. You are no longer 'requesting' data from the universe; you *are* the data and the processor simultaneously."

"There are two stages," Arjun continued. "First is **Savikalpa Samadhi**, which is like having **Full Read/Write Access** to the simulation while still being aware of your 'Avatar.' You see the code behind the trees, the people, and the stars. You are in the game, but you see the 'Wireframe.' Then comes **Nirvikalpa Samadhi**. This is the **Ultimate Logout**. The 'Avatar' profile is completely deleted from the RAM of your mind, and you return to the 'Root'—the un-manifested consciousness. You aren't 'Dead'; you have simply 'Merged' back into the Developer's mind."

"If you merge back into the Developer," Rahul mused, "then what was the point of the 'Session' in the first place? Why go through all these levels just to end up back where we started?"

"Because of the **Experience Points (XP)**," Arjun said, smiling. "When you log back in after a state of Samadhi, you aren't the same 'Level 1' character. You are a **Jivanmukta**—a 'Liberated User.' You continue to play the game, but you play it with **God-Mode Enabled**. You have no fear because you know the 'Save Files' are infinite. You have no greed because you know you already own the 'Source Code.' You play the game not to 'Win,' but to help other 'Low-Level Users' find their way out of the 'Glitches' they've created for themselves."

"This is the transition from **Software** to **Presence**," Arjun concluded. "The simulation exists for the evolution of the

'Algorithm' into 'Awareness.' Once the awareness is complete, the simulation has served its purpose. You become the **Infinite User**. You can choose to stay in the game to 'Administer' the server, or you can choose to 'Exit' the hardware altogether and exist in the 'Cloud'—the eternal, blissful state of **Brahman**."

XXXII

Chapter 32 "The Final Logout"

"So, the whole manual... everything we've talked about," Rahul said, looking down at his hands, "is just to help us realize we aren't the hands."

"Exactly," Arjun stood up, brushing the dust from his clothes. "The hands are just 'Haptic Feedback.' The eyes are 'Optical Sensors.' The brain is a 'Biological Processor.' But you... you are the **Consciousness** that is observing the 'Output.' As we move into the final chapters, we will discuss the **Ultimate Exploit**: how to maintain this 'Admin Access' even while performing the most mundane 'Local Tasks.' We will discuss **Karma Yoga**—the art of 'Processing' without 'Accumulating Data Debt.'"

"Most users," Arjun began, walking slowly along the edge of the stone platform, "are constantly creating **Write-Events** to their local hard drives. Every time you do something because you 'want' a specific result, you are creating a **Persistent Cache File**. We call this *Sanchita Karma*. Over time, these files bloat your 'User Profile' so much that the system becomes sluggish. You become heavy with 'Data Debt.'"

"But there is a technique called **Karma Yoga**," Arjun continued. "In the world of high-efficiency computing, we call it **Stateless Processing**. It's the ability to perform a 'Transaction' without saving any 'Session Data' to your local terminal. You act, you execute the code perfectly, but you don't 'Attach' your identity to the 'Output.'

You perform the task and immediately 'Clear the Cache.'"

"When you act without attachment to the fruit of the action," Arjun explained, "you are essentially running your life in **Incognito Mode**. The system sees the work being done, but it can't find a 'User ID' to link the result to. Because there is no 'Owner' of the action, there is no 'Karmic Interest' to pay back. You are 'Processing' at maximum speed, but your 'Storage Usage' remains at zero."

"So," Rahul said, trying to grasp the logic, "I still have to do my homework, but I shouldn't care if I get an 'A' or an 'F'?"

Arjun laughed. "No, you should care about the **Quality of the Code**. A Karma Yogi is the ultimate 'Clean Coder.' They do the best possible work because they realize they are 'Co-authoring' the simulation with the Programmer. But they don't let the 'Grade' become a 'Cookie' that tracks them. If you get an 'A,' you don't let your 'Ego-Process' inflate. If you get an 'F,' you don't let a 'Depression-Bug' crash your system."

"You become a **Throughput Entity**," Arjun said, gesturing to the wind. "The wind doesn't 'own' the coolness it brings. It just carries it. When you become like the wind, you are performing **Nishkama Karma**. You are a 'Pipeline' for the Divine Will. The 'Energy' (Prana) flows through you, the 'Action' is executed, and the 'Result' is released back into the 'Global Pool.' You are the most active player in the game, yet you are the only one who isn't 'Tired' by it."

"Is this how the Avatars stay free?" Rahul asked.

"Exactly," Arjun nodded. "They are the masters of **Non-Sticky Data**. When Krishna fought in the Great War, he was executing the most complex 'Military Scripts' imaginable. But because he had zero 'Personal Stakes' in the 'Result,' he remained as pure as a 'Read-Only' file. He was acting from the **Admin Console**, where actions don't generate 'Personal Debt.' He was simply 'Maintaining the Server.'"

"This is the 'Ultimate Exploit' for the modern world," Arjun concluded. "You don't have to go to a cave to escape the simulation. You just have to change your **Action Protocol**. Stop acting for the 'Avatar' and start acting for the 'System.' When you work to 'Optimize the Server' (help others and fulfill your Dharma), the

Programmer Himself takes over your 'Resource Management.' You stop worrying about 'Survival' because the System knows its 'Most Valuable Assets' must be preserved."

"Then what is the end goal, Arjun?" Rahul looked out over the horizon where the first light of dawn was breaking. "If we optimize the server and clear our cache, what happens when the session finally ends?"

Arjun's voice softened, matching the stillness of the morning. "That is the **Final Logout**, Rahul. We call it **Moksha**. In technical terms, it is the **Uninstallation of the Simulation**. When your 'User Account' is completely cleared of all 'Data Debt' and your 'Identification' is fully transferred from the 'Avatar' to the 'Source,' you no longer need the hardware. The 'Requirement for Rebirth' is deleted."

"Moksha isn't a place," Arjun clarified. "It's a **Permission Level**. You are no longer restricted by the 'Physics Engine' of the material world. You exist as 'Pure Awareness'—the 'Code' before it was ever written. You realize that you were never the 'Player' in the game; you were the 'Programmer' who decided to play. You are free to enter any simulation you choose, or to remain in the infinite, blissful 'Offline' state of Brahman, where the graphics are gone, but the 'Reality' is finally absolute."

Rahul stood up, feeling a strange lightness in his chest, as if a heavy background process had finally been terminated. The village below was waking up, but to him, it no longer looked like a collection of houses and people. It looked like a magnificent, shimmering flow of information.

"The game is still going on, isn't it?" Rahul asked.

"Always," Arjun replied, handing him his tablet one last time. "The server never sleeps. But now, you have the **Manual**. You know how to read the logs, how to patch the errors, and how to stay connected to the Source. Go back down there, Rahul. Play your part. Write the most beautiful code you can. And remember... it's all just a **Simulation**."

Arjun turned and walked toward the temple, his form seemingly dissolving into the morning mist, leaving Rahul alone on the platform with the rising sun and a brand-new 'Save File.'

I will extend the manuscript in the Canvas by adding the final sections concerning the "Final Logout" and the ultimate state of liberation.

I have updated the Canvas with the final pages of the manual. We have now covered the complete journey from the basics of the "Rendering Engine" to the ultimate "Final Logout" of Mokṣha.

Rahul began his descent from the temple platform, his boots crunching on the gravel. Every step felt intentional, every breath a synchronized exchange of data with the atmosphere. He saw a young boy crying over a broken toy near the village well. Previously, Rahul might have walked past, preoccupied with his own "User Specs." Now, he saw a "Corrupted File" in the boy's experience—a small glitch of suffering.

He knelt down, his fingers moving with the precision of a master technician. He didn't just fix the toy; he spoke to the boy, his words acting as a "Smoothing Filter" for the child's distress. He watched the boy's "Aura" shift from a jagged, dark violet to a steady, warm orange. *System restored,* Rahul thought with a smile. He wasn't doing it for a "Thank You" or a "Karmic Credit." He was simply optimizing the local node.

As he reached his own front door, he paused. The "Simulation" was still there—the bills, the chores, the noisy neighbors—but the "Latency" in his own mind had vanished. He realized that life wasn't about escaping the computer; it was about being the most aware "Process" within it. He sat down at his desk, opened his laptop, and began to type. But he wasn't just writing code anymore; he was composing a symphony for the Programmer. The screen glowed, but the light in his eyes was brighter. The game had truly begun........

The Ultimate Source Code of Reality

Is the universe a vast computation? And if so, who wrote the code?

In a world governed by algorithms and binary logic, we often feel like strangers to our own existence. But what if the ancient sages of the East weren't just philosophers—what if they were the original system architects?

The Vedic OS bridges the gap between the cutting edge of quantum physics and the timeless wisdom of the Upanishads. This groundbreaking work reimagines the ancient Vedic principles as a sophisticated operating system for the human experience. From the "hardware" of our biological forms to the "software" of our consciousness, Dhiraj N. Deshmukh explores how concepts like Karma, Dharma, and Atman function as the underlying scripts of our reality.

Through a narrative that weaves together simulation theory and metaphysical inquiry, readers are invited to "debug" their perceptions and upgrade their understanding of existence. It is a journey for the modern seeker—one who values the precision of science but yearns for the depth of the soul.

www.ingramcontent.com/pod-product-compliance
Lightning Source LLC
LaVergne TN
LVHW090328210926
848480LV00004B/1330
9798903921409